World Scientific Lecture Notes in Physics – Vol. 79

Geometry and Phase Transitions in Colloids and Polymers

World Scientific Lecture Notes in Physics

Published titles

Vol. 60: Massive Neutrinos in Physics and Astrophysics (2nd ed.)
R N Mohapatra and P B Pal

Vol. 61: Modern Differential Geometry for Physicists (2nd ed.)
C J Isham

Vol. 62: ITEP Lectures on Particle Physics and Field Theory (in 2 Volumes)
M A Shifman

Vol. 64: Fluctuations and Localization in Mesoscopic Electron Systems
M Janssen

Vol. 65: Universal Fluctuations: The Phenomenology of Hadronic Matter
R Botet and M Ploszajczak

Vol. 66: Microcanonical Thermodynamics: Phase Transitions in "Small" Systems
D H E Gross

Vol. 67: Quantum Scaling in Many-Body Systems
M A Continentino

Vol. 69: Deparametrization and Path Integral Quantization of Cosmological Models
C Simeone

Vol. 70: Noise Sustained Patterns: Fluctuations and Nonlinearities
Markus Loecher

Vol. 71: The QCD Vacuum, Hadrons and Superdense Matter (2nd ed.)
Edward V Shuryak

Vol. 72: Massive Neutrinos in Physics and Astrophysics (3rd ed.)
R Mohapatra and P B Pal

Vol. 73: The Elementary Process of Bremsstrahlung
W Nakel and E Haug

Vol. 74: Lattice Gauge Theories: An Introduction (3rd ed.)
H J Rothe

Vol. 75: Field Theory: A Path Integral Approach (2nd ed.)
A Das

Vol. 76: Effective Field Approach to Phase Transitions and Some Applications
to Ferroelectrics (2nd ed.)
J A Gonzalo

Vol. 77: Principles of Phase Structures in Particle Physics
H Meyer-Ortmanns and T Reisz

Forthcoming title

Vol. 78: Foundations of Quantum Chromodynamics: An Introduction to Perturbation
Methods in Gauge Theories (Third Edition)
T. Muta

World Scientific Lecture Notes in Physics – Vol. 79

Geometry and Phase Transitions in Colloids and Polymers

William Kung

Northwestern University, USA

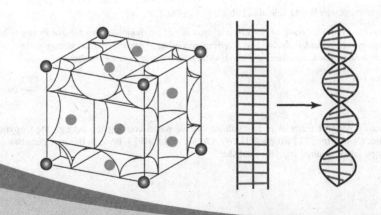

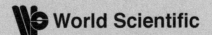 **World Scientific**

NEW JERSEY · LONDON · SINGAPORE · BEIJING · SHANGHAI · HONG KONG · TAIPEI · CHENNAI

Published by

World Scientific Publishing Co. Pte. Ltd.
5 Toh Tuck Link, Singapore 596224
USA office: 27 Warren Street, Suite 401-402, Hackensack, NJ 07601
UK office: 57 Shelton Street, Covent Garden, London WC2H 9HE

Library of Congress Cataloging-in-Publication Data
Kung, William.
    Geometry and phase transitions in colloids and polymers / by William Kung.
      p. cm. -- (World Scientific lecture notes in physics ; v. 79)
    Includes bibliographical references and index.
    ISBN 978-981-283-496-6 (alk. paper)
    1. Phase transformations (Statistical physics) 2. Colloidal crystals. 3. Phase-transfer
catalysts. I. Title.
    QC175.16.P5K86 2009
    530.4'1--dc22

                                                            2009014203

British Library Cataloguing-in-Publication Data
A catalogue record for this book is available from the British Library.

Printed in Singapore.

To David and My Parents

Philosophy is written in this great book (by which I mean the universe) which stands always open to our view, but it cannot be understood unless one first learns how to comprehend the language and interpret the symbols in which it is written, and its symbols are triangles, circles, and other geometric figures, without which it is not humanly possible to comprehend even one word of it; without these one wanders in a dark labyrinth.

Galileo Galilei (1623)

Preface

This monograph represents the culmination of a long journey upon which I first embarked a decade ago in 1997. It was after the first term of my junior year at California Institute of Technology that I realized that my true passion was in theoretical physics and not in chemistry or medicine. I switched to becoming a physics major instead, and I had the good fortune of having John Schwarz as my new undergraduate advisor. Given how I first became fascinated with physics after reading a copy of *Hyperspace* by Michio Kaku way back in high school, I could not wait to jump on the chance then of taking Ph 205 and Ph 239 on Quantum Field Theory and General Relativity as a brand-new, wide-eyed physics major.

My love for independence took me to the University of Pennsylvania on the east coast for graduate school. There I had the good fortune of meeting my graduate thesis advisor Randall Kamien who taught me about beautiful things that abound in "soft" systems. I had a feeling of truly coming in full circle at the end, when I defended my dissertation on geometrical methods in soft-condensed matter physics, the research behind which combined the use of my prior background in biology and chemistry with my love of mathematics and theory.

This monograph represents an extension of my original thesis and includes a more thorough discussion on the concepts and mathematics behind my research works on the foam model, as applied to studying issues of phase stability and elasticity for various non-closed packed structures found in fuzzy and colloidal crystals, as well as on a renormalization-group analysis regarding the critical behavior of loop polymers upon which topological-constraints are imposed. The common thread behind these two research works is their demonstration of the importance and effectiveness of utilizing geometric and topological concepts for modeling and understanding soft systems undergoing phase transitions.

Giving credit where it is due, my thesis advisor Randy Kamien first used the brilliant idea of the foam model in explicating the observed non-close packed structures in fuzzy colloidal systems. Together with Primoz Ziherl, Randy and I extended the foam model in studying elasticity of these fuzzy systems, while at the same time investigated the phase properties in charged colloidal crystals. Most notably, we were all fascinated with the particular A15 lattice structure that first found attention elsewhere in the related mathematical topic of minimal surfaces. It was during my collaboration with Randy when I first appreciated the fact that beautiful ideas from geometry and topology have their natural settings not only in such esoteric subjects as string theory but also in the much more hands-on field of soft physics wherein we can readily see, smell, or even touch the underlying physical systems that we are studying.

For that newfound appreciation as well as the complete freedom afforded me to explore physics and the sciences in general during my years at Penn, I am forever grateful to Randy who is one of the most upstanding and passionate physicists I have met — not to mention being the best science teacher from whom I have had the privilege of learning. Among the other mentors who have deeply shaped me as a scientist during my academic career, I would like to acknowledge my special gratitude to Sunny I. Chan at Caltech for allowing me opportunities to explore my interests in chemistry as well as for giving me numerous sound advice about life when I needed them. I would also like to thank Barry Simon at Caltech for agreeing to be my SURF mentor and granting me a summer of "smelling the roses" in the wild fields of mathematics. I would like to thank Kip Thorne at Caltech for sharing his experience on the importance of ethics in the national and international enterprises of scientific research. In the same vein, I am grateful to my postdoctoral advisor Monica Olvera de la Cruz at Northwestern University for setting a good example in balancing life and science as well as restoring my hope and faith in the latter. Above all, I am indebted to all of them, as well as my dearest friend Alan G. Roche and others too many to mention, for genuinely having my best interest in their hearts.

Last but not least, this monograph also documents, in a way, my growth and progress in this very personal life-learning journey of mine during the past decade. As such, I thank World Scientific Publishing for the opportunity of publishing this research monograph. Of course, I would not be here without my family. And I gladly dedicate this work to my brother David as well as to my parents for their continual and unquestioning support in all my endeavors.

William Kung
Chicago, Illinois
2008

Contents

Preface   vii

List of Figures   xv

List of Tables   xxii

The Big Picture   1

1. Modern Physics at a Glance   3

Geometry and Phase Transitions, in General   15

2. Phase Transitions and Critical Phenomena   17

   2.1   Introduction   17

        2.1.1   Evolution of the Universe: Decoupling of the Four
               Fundamental Forces   18

        2.1.2   Three States of Water   19

        2.1.3   Spins and Magnetism   21

   2.2   Modern Classification of Phase Transitions   23

   2.3   First-Order Phase Transitions: Solid-Liquid Transition . .   24

   2.4   Second-Order Phase Transitions: Scaling and Universality   25

   2.5   Renormalization Group   26

        2.5.1   Kadanoff Picture: Coarse-Graining of Spin Blocks   26

        2.5.2   General Formulation   28

        2.5.3   Critical Exponents   31

        2.5.4   Origin of Universality Class   32

| | 2.5.5 | Wilsonian Picture: Momentum-Space Renormalization Group | 33 |

    2.5.5   Wilsonian Picture: Momentum-Space Renormalization Group 33

  2.6   Mathematical Miscellanies: Semi-Group Structure and Fixed-Point Theorems 34

    2.6.1   Semi-groups 34

    2.6.2   Miscellany on Fixed-Points 35

  2.7   Conclusion 35

3. Overview of Density-Functional Theory   38

  3.1   Introduction 38

  3.2   Electronic Density-Functional Theory 38

  3.3   Classical Density-Functional Theory 42

  3.4   Conclusion 46

4. Survey of Solid Geometry and Topology   49

  4.1   Introduction 49

  4.2   Lattice Symmetry Groups 50

  4.3   Two-Dimensional Space Groups 53

    4.3.1   Hermann-Mauguin Crystallographic Notation . . 55

    4.3.2   Orbifold notation 57

    4.3.3   Why Are There Exactly 17 Wallpaper Groups? . 78

    4.3.4   Other Aspects of Topology in Physics 84

  4.4   Three-Dimensional Point Groups 85

    4.4.1   Face-centered Cubic (FCC) Lattices 85

    4.4.2   Body-Centered Cubic (BCC) Lattices 88

    4.4.3   A15 Lattices 89

  4.5   Conceptual Framework of the Foam Model 90

  4.6   The Kelvin Problem and the Kepler Conjecture 92

  4.7   Conclusion 97

Geometry and Phase Transitions, in Colloidal Crystals   **101**

5. Lattice Free Energy via the Foam Model   103

  5.1   Introduction 103

  5.2   Bulk Free Energy 104

  5.3   Interfacial Free Energy 109

    5.3.1   Charged Colloidal Crystals 109

        5.3.2   Fuzzy Colloidal Crystals 111
    5.4   Conclusion 112

6.   Phases of Charged Colloidal Crystals   115

    6.1   Introduction 115
    6.2   Phase Transitions of Charged Colloids 117
    6.3   Foam Analogy and Charged Colloids 119
    6.4   Conclusion 120

7.   Elasticity of Colloidal Crystals   122

    7.1   Introduction 122
    7.2   Foam Analogy and Cubic Elastic Constants 124
    7.3   Elasticity of Charged Colloidal Crystals 129
    7.4   Elasticity of Fuzzy Colloids 137
    7.5   Conclusion 143

Geometry and Phase Transitions,
in Topologically Constrained Polymers   **145**

8.   Topologically-Constrained Polymers in Theta Solution   147

    8.1   Introduction 147
    8.2 $O(N)$-Symmetric ϕ^6-Theory 148
    8.3   Chern-Simons Theory and Writhe 154
    8.4   One-Loop Scaling of Closed Polymers 159
    8.5   Two-Loop Results 163
    8.6   Conclusion 170

Summary   **175**

9.   Final Thoughts   177

Index   179

List of Figures

2.1 Schematic phase diagram of water. The critical point between the liquid and gas phase represents a second-order phase transition at temperature $T = 647$ K and pressure $P = 2.2 \times 10^8$ Pa. The triple point denotes the condition that all three phases coexist in equilibrium. This only occurs at $T = 273$ K and $P = 6 \times 10^3$ Pa. 20

2.2 Coarse-graining of Spin Blocks: In the Kadanoff construction, we group four spins in neighboring vertices as a unit and consider their average their overall spin value. The resulting "spin-block" constitute a new fundamental degree of freedom after the averaging (coarse-graining) process. Assuming an initial lattice spacing of a between neighboring spins, the rescaled spin blocks have a rescaled spacing of ca, which is increased by a factor of c. 27

4.1 Construction of Wigner-Seitz cells. We first select a reference lattice point (grey circle). We then construct the perpendicular bisectors on all primitive translation vectors (a). The union of the set of all such perpendicular bisectors result in the corresponding Wigner-Seitz cell enclosing the original reference lattice point (b). 51

4.2 Two-dimensional Lattices. Oblique lattice (a). Rectangular lattice (b). Centered rectangular lattice (c). Square lattice (d). Hexagonal lattice (e). 54

4.3 Example of the wallpaper group of $c2mm$ in real life. Common household bricks possess the pattern symmetry of the wallpaper group $c2mm$ in arrangement commonly found in brick walls. (Source: Wikipedia) 56

4.4   Unit cell of *c2mm*. (Source: Wikipedia)   57

4.5   *p1g1*. In orbifold notation, *xx*. The above pattern is constructed from the repeated motif of the letter "k" shown in the lower right corner.   58

4.6   Unit cell of *p1g1*. (Source: Wikipedia)   59

4.7   *p2*. In orbifold notation, 2222. The above pattern is constructed from the repeated motif of the letter "k" shown in the lower right corner. We note that the motif itself has no nontrivial element of internal symmetry other than identity.   61

4.8   *p1*. In orbifold notation, *o*. The above pattern is constructed from the repeated motif of the letter "k" shown in the lower right corner. We note that the motif itself has no nontrivial element of internal symmetry other than identity.   62

4.9   *p1m1*. In orbifold notation, **. The above pattern is constructed from the repeated motif of the letter "k" shown in the lower right corner.   63

4.10   *p2mm*. In orbifold notation, *2222. The above pattern is constructed from the repeated motif of the letter "k" shown in the lower right corner.   64

4.11   *p2mg*. In orbifold notation, 22*. The above pattern is constructed from the repeated motif of the letter "k" shown in the lower right corner.   65

4.12   *p2gg*. In orbifold notation, 22*x*. The above pattern is constructed from the repeated motif of the letter "k" shown in the lower right corner.   66

4.13   *c1m1*. In orbifold notation, **x*. The above pattern is constructed from the repeated motif of the letter "k" shown in the lower right corner.   67

4.14   *c2mm*. In orbifold notation, 2*22. The above pattern is constructed from the repeated motif of the letter "k".   68

4.15   *p4*. In orbifold notation, 442. The above pattern is constructed from the repeated motif of the letter "k".   69

4.16   *p4mm*. In orbifold notation, *442. The above pattern is constructed from the repeated motif of the letter "k".   70

4.17   *p4gm*. In orbifold notation, 4*2. The above pattern is constructed from the repeated motif of the letter "k".   71

4.18   *p3*. In orbifold notation, 333. The above pattern is constructed from the repeated motif of the letter "k".   72

4.19 $p3m1$. In orbifold notation, *333. The above pattern is constructed from the repeated motif of the letter "k".   73

4.20 $p31m$. In orbifold notation, 3*3. The above pattern is constructed from the repeated motif of the letter "k".   74

4.21 $p6$. In orbifold notation, 632. The above pattern is constructed from the repeated motif of the letter "k".   75

4.22 $p6mm$. In orbifold notation, *632. The above pattern is constructed from the repeated motif of the letter "k".   76

4.23 Construction of cylinder (a), Möbius strip (b), and torus (c) based on different identifications (equivalence relations) on the edges of a square disc, parametrized by $\{(x + 2\pi n_x, y + 2\pi n_y) \in \mathbb{R}^2 | n_x, n_t \in \mathbb{Z}\}$.   79

4.24 Euler characteristics for different surfaces. Dodecahedron, $\chi = 2$ (a). Sphere, $\chi = 2$ (b). Möbius strip, $\chi = 0$ (c). Torus, $\chi = 0$ (d). Double torus, $\chi = -2$ (e). Klein's bottle, $\chi = 0$ (f). Triple torus, $\chi = -4$ (g). Figures are obtained using Mathematica (http://www.wolfram.com/mathematica).   81

4.25 Examples of simplexes. A simple geometric point corresponds to a 0-simplex (p_0). A line corresponds to a 1-simplex $\langle p_0 p_1 \rangle$. The mathematical object of a 2-simplex is denoted by $\langle p_0 p_1 p_2 \rangle$. The tetrahedron is an example of a 3-simplex denoted by $\langle p_0 p_1 p_2 p_3 \rangle$.   83

4.26 The 14 Bravais lattices in three dimensions. (a) Simple cubic [P]. (b) Body-centered cubic [I]. (c) Face-centered cubic [F]. (d) Simple tetragonal [P]. (e) Body-centered tetragonal [I]. (f) Orthorhombic [Γ]. (g) Orthorhombic [C]. (h) Orthorhombic [I]. (i) Orthorhombic [F]. (j) Monoclinic [P]. (k) Monoclinic [C]. (l) Trigonal [R]. (m). Hexagonal [P]. (n) Triclinic [P]. . . .   86

4.27 Unit cell of the FCC lattice (a) and the corresponding three-dimensional Wigner-Seitz cells with the geometry of rhombic dodecahedron (b). The net image of a unit cell of the rhombic dodecahedron (c). (Source: Kamien group, University of Pennsylvania, Philadelphia.)   87

4.28 Unit cell of the BCC lattice (a) and the corresponding three-dimensional Wigner-Seitz cells with the geometry of Kelvin's tetrakaidecahedron (b). The net image of a unit cell of the Kelvin's tetrakaidecadedron (c). (Source: Kamien group, University of Pennsylvania, Philadelphia.)   89

4.29 Unit cell of the A15 lattice (a) and the corresponding three-dimensional Wigner-Seitz cells with the geometry of Weaire-Phelan minimal surface (b). (Source: Kamien group, University of Pennsylvania, Philadelphia.)   92

4.30 A two-dimensional illustration of the minimal-area rule: the hard cores are embedded in the matrix of coronas, and the volume of the latter is given by the product of its area (dotted line) and the average separation of the cores.   93

4.31 Truncated Octehedron. The geometric structure that solves Kelvin's conjecture in finding the ideal configuration of a soap froth. Figures are obtained using Mathematica (http://www.wolfram.com/mathematica).   96

4.32 Measurements of the square pyramid truncated from the octahedron in orthic tetrakaidecahedron; $d = \frac{\sqrt{2}}{2}a$.   97

5.1 Various lattices: (a) face-centered cubic, (b) body-centered cubic, and (c) A15 lattices.   106

5.2 Rhombic dodecahedron/FCC lattice (a), Kelvin's tetrakaidecahedron/BCC lattice (b), and Weaire-Phelan minimal surface/A15 lattice (c).   107

5.3 Approximate analytical model of the bulk free energy for the A15 lattice: free volumes of columnar and interstitial sites are replaced by cylinders and spheres, respectively.   108

5.4 Bulk free energies of hard-core particles arranged in FCC, BCC, and A15 lattice as calculated with the free-volume theory. Solid lines correspond to Eqs. (5.4) and (5.5), and circles are numerical results.   109

5.5 Interactions between fuzzy colloids. For colloidal particles of the fuzzy kind, there are two main types of interactions: hardcore interaction stemming from the non-penetrability between the particle cores; soft-tail interactions between neighboring particles that increase with decreasing interparticle distances.   112

6.1 Theoretical FCC-BCC coexistence curve as a function of volume fraction ϕ and electrolyte concentration (HCl). The diamonds are the theoretical predictions, while the other points come from the data of Ref. [21]. Solid squares are coexistence points, open triangles are FCC, and open circles are BCC.   119

7.1   The two different types of deformations: (a) elongational shear
       mode; shear deformation along the four-fold axis (b) simple
       shear mode; shear deformation along the face diagonal. In
       both cases, we only consider deformations which preserve the
       total volume of the unit cell. The elongational shear mode is
       parametrized by q, while the simple shear mode is parametrized
       by s.   126

7.2   The bulk free energy of the BCC lattice as a function of the
       elongational shear at $n = 1.1$. The thick line represents the
       quadratic fit to the curve at $q = 0$ which determines the bulk
       part of the shear modulus. The minimum at $q \approx 0.26$ corre-
       sponds to the FCC lattice.   127

7.3   The surface area parameter γ as function of dimensionless pa-
       rameter q for the elongational shear mode. The kinks in the
       curves indicate a change in the topology of the unit cell.   128

7.4   The bulk modulus K for the FCC, BCC, and A15 lattices of
       a charged colloidal system as a function of $\lambda = \kappa a$, where a is
       the average interparticle spacing. The calculations are done at
       density $n = 0.9$ and at dimensionless surface potential $\Phi_s = 0.4$.
       The maximum at $\lambda = 3$ is spurious and signals the breakdown
       of our approximation (see text).   131

7.5   The elastic constants K_{11}, K_{12}, and K_{44} for the FCC, BCC,
       and A15 lattices of a charged colloidal system as a function of
       $\lambda = \kappa a$, at density $n = 0.9$ and at dimensionless surface potential
       $\Phi_s = 0.4$.   132

7.6   The bulk modulus K for the FCC, BCC, and A15 lattices of
       a charged colloidal system as a function of density n, at fixed
       screening length $\lambda = 4$ and at surface potential $\Phi_s = 0.4$. The
       bulk modulus K increases with increasing density n for all three
       lattices and diverges upon close-packing.   133

7.7   The elastic constants K_{11}, K_{12}, and K_{44} for the FCC, BCC, and
       A15 lattices of a charged colloidal system as a function of density
       n, at fixed $\lambda = 4$ and at surface potential $\Phi_s = 0.4$. Similar to
       the bulk modulus K, the moduli of the elastic constants diverge
       upon close packing.   134

7.8   (a) Schematic of the shear instability of the A15 lattice: the cubic arrangement of the spheres (solid line) is unstable to shear along the face diagonal (dashed line). (b) The cubic A15 lattice and the triclinic derivative lattice as the entropically preferred structure at high densities.   136

7.9   The effective dimensionless "screening length" of the dendrimers, $L = \ell/\sigma$, corresponds to the ratio of the thickness of the soft outer part, consisting mainly of the alkyl corona, to the radius of the impenetrable hard core of the dendrimer molecule consisting of rigid aromatic rings.   137

7.10 The bulk modulus for the FCC, BCC, and A15 lattices of a fuzzy colloidal system as a function of "screening" length $L = \ell/\sigma$, at fixed density $n = 0.9$. As the length of the corona increases with increasing L, the thermodynamically favorable structure transits from FCC, to BCC and eventually to the A15 lattice at high enough L. The corresponding bulk modulus simply behaves linearly in the respective regions.   138

7.11 The elastic constants K_{11}, K_{12}, and K_{44} for the FCC, BCC, and A15 lattices of a fuzzy colloidal system as a function of "screening" length $L = \ell/\sigma$, at fixed density $n = 0.9$. As the "screening" length L increases, non-close packed structures are favored. Similar to the bulk modulus, the three elastic constants behave linearly with respect to L.   139

7.12 The bulk modulus K for the FCC, BCC, and A15 lattices of a fuzzy colloidal system as a function of density n, at "screening" length $L = \ell/\sigma = 0.05$. As expected, the bulk modulus K increases with increasing density n.   139

7.13 The elastic constants K_{11}, K_{12}, and K_{44} for the FCC, BCC, and A15 lattices of a fuzzy colloidal system as a function of density n, at "screening" length $L = \ell/\sigma = 0.05$. They increase with increasing density n much like the bulk modulus.   140

7.14 The shear constants $\mu = K_{11} - K_{12}/2$ for the FCC, BCC, and A15 lattices of a fuzzy colloidal system as a function of density n and "screening" length $L = \ell/\sigma$. They result from the elongational shear mode and represent the lower bound of the average shear measurable in experiments.   141

7.15 The shear constants $\mu = K_{11} - K_{12}/2$ for the FCC, BCC, and A15 lattices of a charged colloidal system as a function of density n and $\lambda = \kappa a$. They increase with increasing density, as expected, and are the theoretical lower bound to the range of average shear. 142

8.1 Phase Diagram for a Tricritical System. The line $\mu = 0$, $\lambda > 0$ is a second-order lambda line, shown as a single line in the figure. The line $\mu = \frac{1}{2}|\lambda|^2/\nu$ is a line of first-order transition, shown as a thickened grey line in the figure. The triple point occurs at $\mu = 0$, $\lambda = 0$. 150

8.2 Some possible paths for the high-temperature expansion of the Ising model for spins on a square lattice. While the path $(i \rightarrow j)$ is self-avoiding, the path $(k \rightarrow l \rightarrow m \rightarrow n)$ is self-intersecting and vanishes in the $N \rightarrow 0$ limit. 152

8.3 First-order corrections to Z_ϕ (a), Z_w (b), and $Z_\mu(p = 0)$ (c) in $O(N)$-symmetric ϕ^6-theory. Note that figures (a) and (b) are already four loops in contribution. 154

8.4 Twist can be understood as the rotation of cross section about the tangent of the two backbones; the overall length remains constant in this mode of topological constraint (a); Writhe is simply the integrated torsion of a line; unlike twist, the projected length of the object changes in this mode (b). 155

8.5 One-loop diagrams arising from the gauge field coupling to matter: correction to the gauge field self-energy (a), (b); correction to matter self-energy (c), (d); correction to the three-point vertex (e), (f). 161

8.6 One-loop contributions to $\Gamma^{(6,0)}$. 162

8.7 Two-loop diagrams arising from the gauge field coupling to the scalar field: contributions to Z_ϕ (a), (b); corrections to the cubic gauge vertex (c)–(f); contributions to $Z_\mu(p = 0)$ (g). 165

8.8 Two-loop contributions to $\Gamma^{(6,0)}$. Figures (c), (d) and (e) are representative of the topology; we have accounted for other contractions of ϕ with ϕ^*. 166

List of Tables

7.1 Calculated values for the bulk modulus K and elastic constants of the BCC structure at several values of $\lambda = \kappa a$, where a is the average interparticle spacing, for $n = 0.23$ (first number) and $n = 0.50$ (second number). All values are in units of N/m^2. . .   130

8.1 Divergent contribution from each graph and to the appropriate renormalization constant. Note that there are no divergences at one loop.   168

PART 1
The Big Picture

Chapter 1

Modern Physics at a Glance

The progress of humankind has always been in parallel with our improved understanding and the consequent mastery of the natural elements around us. In particular, our ability to engineer natural materials into useful tools and devices of desirable properties — be they mechanical, electrical, or chemical — is paramount to the advancement of our own civilization. This tandem march towards overall societal progress is so recognizably intrinsic that early milestones of human civilization are named after the characteristic natural materials that were mastered by our progenitors during the specific time periods. For instance, the stone age and the bronze age have been so named for our ancestral achievement in mastering the tool-making techniques based on stone and bronze, respectively, for the first time in history. Fast forwarding to modern times, the industrial revolution would not have been possible without the tremendous progress in our firm grasp of metallurgy, of the principles of *thermodynamics* that spawned the steam engine and later the internal combustion engine, and of *electromagnetism* that preceded electrical power regeneration — in conjunction with all other social factors at the time including the institutional movement from feudalism to imperialism in Europe as well as the creation of financial markets. Of course, we are presently in the midst of a revolution of our times in information technology. The ramifications thereof are still wide open to determination. Of the most important progeny of our current IT revolution is the World Wide Web, which owes its existence largely to our technical know-hows of semiconductors making use of such materials as silicon, germanium, and gallium arsenide. In this instance, our understanding of semiconductors stems from the fundamental theory of *quantum mechanics*.

Thus, we see that in the prior examples that with each breakthrough in our understanding of materials and matter comes the corresponding step forward in the quality of our daily life. Historically, new understanding of materials and matter has been driven by novel experimental results requiring new theoretical explanations as well as new directions taken by theorists in modeling and predicting properties. In what follows, we will define a *theoretical model* generally as an abstraction of aspects of a physical system for which we are interested in explaining their inner working and properties. Some theoretical models yield predictions that are so encompassing in scope and revolutionary in paradigm that we call them *fundamental theories*. The aforementioned example of quantum mechanics would be one such instance: with the advent of quantum mechanics at the turn of the twentieth century, our understanding of matter and materials advanced to an unprecedented level. As a result of the revolutionary thinking made by the likes of Max Planck, Neil Bohr, Louis de Broglie, Max Born, Werner Heisenberg, Erwin Schrödinger, Wolfgang Pauli, Paul Dirac and others, chemistry has since progressively turned from a field of empirical science to one whose basis can be, in principle, rigorously understood in terms of the quantum physics of atoms and molecules.

It turns out that the discovery of quantum mechanics is only the tip of an iceberg. When we further consider interaction of matter with light (photons) that underlies the foundation of spectroscopy, the theory of *quantum electrodynamics* (QED) comes into play, which merges non-relativistic quantum mechanics with special relativity and electromagnetism. As we probe further into the finer structure of the atom and its nucleus, we reach the domain of *quantum chromodynamics* (QCD) which describes the strong interaction binding the quarks and gluons together that make up the protons and neutrons inside the nucleus of an atom, as well as other types of hadrons. Together with the unified description in terms of the *electroweak theory* of electromagnetism and the weak nuclear force, the latter of which is responsible for radioactivity, we have outlined the *standard model* which underlies the foundation of modern physics. In principle, these theories should encode all information pertaining to all systems to which they are applicable, or equivalently, we should be able to derive the fundamental properties of all such systems from these theories. However, the complexity of such calculations involved often precludes such possibility at the practical level.

Consequently, numerical methods and approximate analytical models continue to be of value in our quest for further understanding of many physical systems in nature, in spite of the already available "cracked codes" — in the guise of the aforementioned fundamental theories — to the inner working of Nature firmly in our hands. In what follows, we will define an *analytical* model as a theoretical model that allows for the explicit evaluation of the relevant quantities pertaining to the physical properties of a system in closed analytical forms. In particular, to distinguish from fundamental theories, we refer to an *approximate* model whose main purpose is to serve as an effective description of the phenomenology of a system in certain useful or interesting limits. These approximate models are also sometimes referred to as *phenomenological models* or *effective models* to emphasize their very specific range of applicability.

On the other hand, *numerical methods* to theoretical modeling comprises analytical models whose predictions of relevant physical parameters can only be obtained through computational algorithms or expressible in numerics. *Simulations* would by construction fall into this category of numerical modeling. Numerical methods are essential as there are only very few classes of problems, pertaining mostly to highly idealized systems, which can be directly cast and solved entirely within the tight constraints imposed by *classical analysis* in mathematics. In fact, both numerical methods and approximate theoretical models are necessary in scientific research due to the computational complexity and often the resulting practical impossibility to deriving everything from fundamental theories; the pursuit of the continual refinement of these numerical methods as well as of approximate analytical models in understanding Nature comprises the bulk activity of the present-day enterprise of scientific research.

As a concrete illustration of the necessity of approximate models and numerical methods in science, we point out the fact that the issue of computational difficulty is already present at the level of *classical mechanics* in the study of three-body systems. This seemingly innocuous three-body problem is already not amenable to a closed-form analytic solution. More generally, for the problem of an N-body system, one can only resort to numerics to plot out the trajectories of motion carried out by the system when the total number of bodies exceeds three, notwithstanding the facts that Newton's laws of motion and gravity are well founded and understood in the language of classical differential equations as well as that the formulation of the N-body problem can be straightforwardly presented in analytical form. With limitations, the existence of an analytical solution in the form of a slowly converging series was eventually established by the Sundman theorem for the three-body problem, but the complexity involved in extending this result to the general N-body problem exemplifies the unfortunate fact that elegance and generality do not always go in tandem in analysis.

Consequently, given the complexity found in real systems, the reductionist approach of the physical sciences, wherein we abstract what we deem as the defining features of a system and strip down to its bareboned ideal analogue that still preserves these features, is a rather necessary and pragmatic philosophy. The prominent roles of numerical methods and approximate models in modern science demonstrate the constant struggle between practical idealism and the seemingly impenetrable reality. Striking a compromise between the two realities, one can go about it in two ways: firstly, to invent more-efficient algorithms in obtaining numerical solutions that would allow us to model, by computational brute force, a more accurate abstraction of the real systems that must necessarily include more complex details; alternatively, one can develop better approximate analytical models that would allow for more accurate or detailed predictions using existing computational tools.

What precisely makes for an improved model over established ones now warrants further attention. While the qualifier for a more efficient algorithm is unequivocal — one that does the same job in less time and/or using less computational resource — the criteria for improved approximate analytical models merit broader consideration. At the most obvious level, an improved approximate model would, as described earlier, provide a new and more efficient computational route for theoretical scientists to produce falsifiable results that can be verified against experimental findings, just like its numerical counterpart. Achieving this goal requires the insights from scientists in discerning which subset of the underlying degrees of freedom for a system contributes most crucially to certain observed properties for the system.

Consider the Brownian motion of a particle in a fluid with constant friction. The erratic trajectory of the Brownian particle is due to the collisions it makes with other fluid particles. There are two dual descriptions of this problem. The first approach is that of the Langevin equation, which is an example of stochastic differential equation. The Langevin equation for this Brownian particle is simply the Newton's equation, with the addition of a source term encapsulating the overall effect of constant collisions that describes the time evolution the particle's position and velocity. On the other hand, the approach of Fokker-Planck equation, which delineates the velocity probability function for the diffusing particle (the generalization to the probability function for displacement and other variables is called the Smoluchowski equation), focuses on a different set of quantities and provides an alternative description to Langevin equation. While one can utilize the Langevin equation to derive the many interesting results of the system without a direct solution (by means of the fluctuation-dissipation theorem), obtaining the direction solution is itself computationally intensive; the Fokker-Planck equation is often used instead. Thus, whichever model would work better depends, in general, on what one's main goals and interests are for the system under study. Of course, these dual descriptions are possible due in large part to their phenomenological nature satisfying a few rather general assumptions about the underlying dynamical systems not far from equilbrium. In addition, we note that numerical methods also exist for this problem with the use of Monte Carlo simulations.

In addition to seeking a minimal set of degrees of freedom that would most directly characterize a desired set of properties for a physical system, another useful approach in constructing good approximate analytical models involves smartly integrating out degrees of freedom that are nonessential; we would then consider the resulting *effective* or *coarse-grained* models, the latter of which refers specifically to models with microscopic degrees of freedom integrated out. For example, *thermodynamics* provides a description for the equilibrium states of systems that contain many degrees of freedom in terms of a few macroscopic system variables such as temperature, internal energy, or magnetization. These macroscopic quantities are, in general, much more useful characterization of a thermal equilibrium system than the knowledge of all colliding trajectories or all microstates associated with either the underlying particles or elementary excitations of the system. When we consider small perturbations to this ideal homogeneous equilibrium at all point in space, in the long-wavelength and low-frequency limit, we turn to the framework of *hydrodynamics* in terms of, again, a given set of macroscopic field variables. In this regard, both thermodynamics and hydrodynamics constitute coarse-grained theories concerning the static and dynamical properties, respectively, of the underlying physical systems with microscopic constituents.

In this work, we will mainly be concerned with the static aspects of two particular condensed-matter systems: colloidal crystals and loop polymers. By *statics*, we refer to such system properties as phase stability and other properties that do not have general time-dependence. In contrast, we refer to *dynamics* the study of all time- or frequency-dependent phenomena such as kinetics and coarsening not far from equilibrium. The main focus of this work is on the phenomena of phase transitions in colloidal crystals as well as in polymers that are subjected to the constraints of non-trivial topology such as loops and knots. We will make connections with various ideas from geometry and topology to derive a new theoretical model in understanding the first-order colloidal phase transitions between the different lattices, in particular the exotic $A15$ lattice that has defied reasonable explanations thus far based on conventional wisdom. We will also make connections with geometry and topology in performing renormalization-group analysis on the second-order polymeric phase transitions and discovering a new scaling regime and new universality class for topologically constrained polymers.

In general, colloids can be classified according to their microscopic interactions between constituents. The intermolecular potential can either be long-range or short-range in nature. Charged colloids are the prototypical colloidal system with long-range constituent interaction: the particles interact via the screened-Coulomb potential, which is also known as the *Yukawa potential*. Previously, many experimental studies provided a wealth of data in regard to the stability of various phases, notably the disordered phase, the face-centered cubic phase (FCC) and the body-centered cubic phase (BCC). These experimental systems consisted typically of aqueous suspension of uniform, charged-polystyrene spheres with variable salt concentration, the latter being a control parameter for the degree of screening of the underlying electrostatic Coulomb interaction.

For prototypical colloidal systems with short-range interactions, we consider colloids of the *fuzzy* kind. Fuzzy systems are made of molecules with long, flexible alkyl chains attached to centers of rigid, aromatic rings. The interparticle potential for these systems may be approximated fairly well by a simple hardcore dressed with a repulsive short-range interaction of finite strength.

On the other hand, polymers are large molecules consisting of repeating chemical units. Each repeating unit, known as monomers, is typically made of more than five and fewer than 500 atoms, while a polymer is qualified by having more than 500 monomers linked together in various topologies — the most familiar one being long-chained linear polymers. Polymers have important industrial applications such as plastics and drug delivery and constitutes the basis of molecular biology.

In probing the stability of various colloidal phases, molecular dynamics (MD) simulations are generally used in addition to experiments. Qualitative corroborations exist between MD and experimental findings. Though it is encouraging that such qualitative agreement exists, one is hard-pressed for similar quantitative convergence between existing experimental data and simulation results. Further complicating the matter is the lack of analytical calculations from which either experimental or simulation findings can be definitively derived. Methods like the *density-functional theory* does not provide an intuitive understanding of the phase behaviors and other thermodynamical properties of these colloidal systems. This difficulty thus strongly necessitates formulation of theoretical models that are not only computationally simple enough to yield insightful solutions but also general enough to have great applicability to a range of diverse systems.

Theoretically, the free energy of a system completely determines its equilibrium thermodynamics. The stable phases of a physical system for a given range of parameters correspond to the minima of its free energy for the respective range. This is in accordance with the *de facto* standard in modern physics that the rigorous formulation of a problem should involve the extremization of some relevant quantities with respect to certain set of constraints that characterize the system. This fact, along with another fact that physics is mostly about the study of *structures* of various systems (the universe, spin lattices, superconductors, *etc.*) might together explain the unusual effectiveness of mathematics as a language in describing the material world, as encapsulated by the *Wigner's Puzzle*.

At the mean-field level where we do not take into account the effects of correlation between the various components or constituents within a physical system, the Landau theory has proven remarkable success in modeling the phenomenology of a variety of systems near transitions between phases, under very simple assumptions based on the symmetry of order parameters. In this regard, Landau theory stands out as one of the most remarkable examples of coarse-grained models that we have first defined earlier.

To further elaborate on the details of Landau theory and coarse-graining, we return to discussing some of the general features of coarse-grained models, now with the specific context of thermodynamics of colloidal and polymeric systems in mind. The rigorous bridge between the microscopic degrees of freedom and the macroscopic field variables is formally established by the subject of *statistical mechanics*, wherein statistical methods are often utilized to extract useful information and characterization of systems approaching the *thermodynamic limit*, in which the total number of degrees of freedom, N, of these systems approach infinity ($N \to \infty$). Within statistical mechanics, the physical concept of coarse-graining has been made rigorous by the mathematics of *renormalization group*, which is perhaps one of the deepest and most beautiful insights arising from modern physics. The theory of renormalization group relates effective models of different scales (such as length, momentum, or energy) in terms of how various interaction parameters and correlation functions change or *renormalize* with respect to these scales. It turns out that the relations of how these interaction parameters, coupling constants, and correlation functions scale with respect to length or energy (renormalization group flows) actually reveal unexpected and deep physics between the different phenomenological models, such as the origin of *universality classes* in *critical phenomena*.

Mathematically, phase transitions occur when non-analyticity develops for the free energy of a system for some choice of thermodynamic variables. It is shown in statistical mechanics that this non-analyticity is only a consequence of taking the system to its thermodynamic limit. Historically, phase transitions were classified according to the *Ehrenfest classification*: the various solid/liquid/gas transitions are deemed *first-order transitions* because they exhibit a discontinuity in the first derivative of the free energy with respect to a thermodynamic variable. On the other hand, the example of ferromagnetic/paramagnetic phase transition is considered a *second-order transition* because of a discontinuity in the second derivative of its free energy. The Ehrenfest classification is incomplete since it is based on the mean-field understanding of phases. The modern classification relies on whether these phase transitions involve *latent heat* (first-order) or not (continuous or second-order).

In general, second-order phase transitions are easier to study than first-order transitions due to the presence of latent heat in the latter processes. A whole host of machinery has been devised to the investigation of critical phenomena, so called because of their association with critical points in the free-energy description. Critical phenomena can be characterized by *critical exponents*, the index associated with the power law by which certain thermodynamic variables, such as the heat capacity, diverge. Remarkably, the phenomenon of universality is observed for continuous phase transitions across systems of wildly different microscopic physics that nevertheless exhibit the same set of critical exponents, thus reflecting a common characteristic long-distance critical behavior. As such, we classify second-order transitions into different *universality classes* as characterized by different sets of these critical exponents. And the method of renormalization group provides a systematic computational tool for the evaluation of these critical exponents, based upon the physical concept of coarse-graining and scaling, once we specify the free energy expression for a system. Therefore, our main challenge in physics and in this particular work is to come up with reasonable free-energy expressions for various systems of interest that would capably capture the physical features underlying these systems to the level of accuracy that we desire.

Historically, physics has always drawn inspirations from development of mathematics. When it comes to modeling real-life physical systems, such as the construction of a suitably chosen free-energy expression, it is no exception. For example, as a first-order approximation, we often approximate microscopic particles as spheres when the specific properties of a system do not explicitly depend on the shape anisotropy of its constituents. Correspondingly, interactions between these microscopic particles are mapped to interactions between geometrical spheres. For more complicated systems such as liquid-crystalline mesogens whose macroscopic phase properties actually depend crucially on the shape anisotropy found in their inter-molecular interactions, we can idealize the constituent molecules with other appropriately chosen geometrical objects — cylinders in the case of liquid crystals. One of the major pillar of twentieth-century modern physics — Einstein's theory of *general relativity* — arose from Einstein's successful elucidation of gravitation upon combining the rigorous mathematical framework of *non-Riemannian geometry* and the physical idea behind the conceptually elegant *correspondence principle*.

Thus, geometry has played a central role in the development of modern physics. Conventional wisdom dictates that formulation of problems, when rooted in a properly chosen geometrical setting, becomes simple and meaningful. For example, Maxwell's equations and more generally the phenomenon of electromagnetism, in the context of gauge symmetry, is the unique consequence of the abelian $U(1)$-gauge field. Furthermore, the strong and weak nuclear forces found in the standard model of particle physics are but physical consequences of the extended non-abelian gauge groups $SU(2)$ and $SU(3)$, respectively. In addition, such symmetry principle as *general covariance* serves as the prerequisite to all properly formulated fundamental models of Nature and expresses the fact that the physics of any system should be independent of any chosen mathematical description in terms of a particular coordinate system. As the main goals of this work, we use concepts from geometry in studying and understand a class of first-order phase transitions between the different lattice phases (FCC, BCC, and A15) found in colloidal crystals, and we apply geometric and topological concepts to studying a class of second-order phase transitions in closed-loop polymeric systems. In particular, we propose a geometrical way of understanding the thermodynamic and elastic properties of colloidal crystals based on what we term as the *foam model*, and we perform a renormalization-group analysis to confirm a new universality class for long polymers with non-trivial topologies.

This monograph is organized as follows. In Chapter 2, we further pursue the general theory of phase transitions. We will illustrate with examples the pervasiveness of phase transitions as a theme in our understanding of Nature within the paradigm of modern physics. We will provide the common classifications of the different types of phase transitions. In particular, we will survey some of the most important concepts pertaining to both first-order and second-order phase transitions. We will conclude the chapter with an important exposition to the concept of renormalization group and the mathematical background behind the conceptual framework.

In Chapter 3, we will provide an account of density-functional theory, one of the main tools available to soft-condensed matter theorists in studying phase transitions. We will discuss density-functional theory both in its original context of electronic many-body problems as well as its adaptation to the classical study of macroscopic phase transitions of statistical systems.

In Chapter 4, we will survey some of the important concepts from solid geometry and topology pertaining to our main results of the foam model in describing phase stability of various lattice structures as well as our renormalization-group analysis applied towards studying phase transitions associated with closed-loop polymers. We will see on a practical level how symmetry can be used in the classification of a plethora of lattice phases found in two- and three-dimensional spaces. We will demonstrate both the Hermann-Mauguin crystallographic notation and the orbifold notation for labeling the different lattice phases. In particular, we will put emphasis on the three cubic lattice structures: namely, the face-centered cubic (FCC) lattice, the body-centered cubic (BCC) lattice, and the β-Tungsten A15 lattice. Structures with these three types of geometry have been found in colloidal systems, and we will see how the foam model can be used to understand their origins. After presenting the conceptual framework of our foam model, we will conclude the chapter by discussing the closely related mathematical problems of the Kelvin problem and of the Kepler conjecture.

In Chapter 5, we will apply our foam model in the direct construction of various free-energy expressions for the different phases of both charged and fuzzy colloidal crystals, in terms of their geometry and interparticle dynamics. Instead of the usual point approximation, we will consider the opposite limit of mapping these spherical colloidal particles as flat surfaces under appropriate experimental conditions.

In Chapter 6, we will present our main result of reproducing an experimental phase diagram [*Phys. Rev. Lett.* **62**, 1524 (1989)] with only one adjustable parameter of experimentally inaccessible surface-charge density of the colloidal particles. We will also discuss our results in the context of existing knowledge on phase properties regarding charged colloids.

In Chapter 7, we apply our foam analogy to studying elastic properties of colloidal crystals. We will construct the elastic free energy expressions for the FCC, BCC, and A15 lattices in both fuzzy colloids and charged colloids and derive the various elastic constants from first principles. We will present a thorough discussion on our main results, including instabilities of the BCC and A15 lattices with respect to certain shear modes.

In Chapter 8, we study the phase transitions of topologically-constrained polymers. Closed-loop polymers (such as DNA) obey the topological constraint given by Fuller's theorem between the Linking number Lk, twist Tw, and writhe Wr: $Lk = Tw + Wr$. To investigate the scaling behavior of these polymers in solution, the thermal fluctuation of system can be systematically handled by the casting the problem in a field-theoretic language. We will first introduce the Chern-Simons gauge field in order to implement the aforementioned Fuller topological constraint for these polymers. The model of interest will be the abelian Chern-Simons theory. We consider the fluctuation effect on the critical exponent associated with the radius of gyration up to two-loop order via the method of renormalization group analysis. We will demonstrate a new universality class for the class of closed-loop polymers. We will show that the critical exponent characterizing the onset of this phase transition depends on the chemical potential for writhe, and it gives way to a first-order transition induced by fluctuation.

We will conclude our survey of geometry and phase transitions in colloids and polymers with a few final thoughts in Chapter 9.

PART 2

Geometry and Phase Transitions, in General

Chapter 2

Phase Transitions and Critical Phenomena

2.1 Introduction

Phase transitions are ubiquitous in nature. Many interesting phenomena occurring in nature are examples of transitions between the different *phases* of corresponding physical systems. Intuitively, a phase of a physical system is characterized by a relatively uniform set of chemical and/or physical properties.

As an example, the physical system in discussion can be as large as the whole universe, as our universe has been evolving by undergoing a series of phase transitions since the beginning of time as we know it, termed the Big Bang. As another example, an essential ingredient to life on Earth depends crucially on the particular phase properties of a chemical element that we call water, with its two hydrogen atoms and one single oxygen atom. The phase properties in the case of water are manifestations of the collective effects of individual constituent molecules all of the same chemical composition and, consequently, are naturally described in terms of macroscopic variables that are continuous functions of space. Biologically, it has even recently been conjectured that our brain must function near a critical threshold of a second-order phase transition in order to account for its robust and flexible behavior, due to a large abundance of metastable states to choose from at criticality [1, 2].

In contrast to the aforementioned examples thus far that involve systems whose phase-transition descriptions are best done in terms of continuous variables, there are also systems possessing discrete degrees of freedom that undergo phase transitions. One such example would be that of magnetic spins described by the Ising and n-vector models, whose critical properties depend on the dimensionality of the system.

As a brief survey of the important ideas behind the modern theories of phase transitions, we will elaborate further in the three examples below on the phase transitions underlying the evolution of our universe, between the different states of water, and of spins and magnetism as previously mentioned.

2.1.1   Evolution of the Universe: Decoupling of the Four Fundamental Forces

In the modern paradigm of cosmology, we can understand the evolution of our Universe since the moment of the Big Bang as a series of phase transitions. While a sufficiently detailed exposition on this topic is beyond the scope of this book, we do mention a few salient stages of the evolution process that illustrate the pervasiveness of the phenomenon of phase transitions all the way up to the grandest cosmological scale and down to the smallest of time-steps from the beginning of time as we know it.

At time $t = 10^{-43}$ seconds, all four fundamental forces of the universe as we know it — gravity, electromagnetism, strong nuclear force, and weak nuclear force — were believed to be simply different manifestations of a single "unified" force. This period of the evolution of our universe is known as *the Planck epoch*. As the universe progressed from the earliest moment of the Big Bang, the average temperature of the universe dropped, and consequently each of these aforementioned four forces "condensed" from the unified force. The first force to decouple from the unified force is gravity, which marked the end of the Planck epoch. At this point, our modern understanding of this era is still scant.

From the end of the Planck epoch until the time at $t = 10^{-38}$ seconds came what we term as *the GUT epoch*. The acronym *GUT* stands for the grand unified theory. The electromagnetic, strong nuclear, and weak nuclear forces were still unified as a single force at this point. The end of the GUT epoch was marked by the breaking of the strong nuclear force from the electromagnetic and weak nuclear forces. The energy released during this phase transition was the main impetus of the ensuing inflation, which explained the isotropy of the cosmic microwave background as well as the asymptotic flatness of the universe. Also of importance is the genesis of a small excess of matter over antimatter during this epoch.

After the inflationary epoch, the electroweak phase transition occurred at $t = 10^{-10}$ seconds during which the electromagnetic and weak nuclear forces became decoupled via the mechanism of spontaneous symmetry breaking. The vacuum energy of the universe also underwent a phase transition so that the Higgs boson first acquired a vacuum expectation value during this period. Consequently, all other particles acquired their respective masses.

The evolution of the universe continued with the generation of quarks, hadrons, leptons, and eventually atoms and molecules, along with the larger cosmic structures such as stars, planets, galaxies, and more. The structure of the observable universe today is a direct consequence of a series of phase transitions of prior times.

2.1.2 *Three States of Water*

Water, being one of the most familiar everyday compound and also amongst the most essential elements of life, provides a dramatic example of phase transitions between its many different but familiar guises. The three most commonly found *states* of matter for water include the *solid* ice, the ubiquitous *liquid* form, as well as *gas* vapor. A schematic phase diagram is shown in Fig. 2.1.

In the solid region of the phase diagram, water exists as ice. Presently, there are fifteen known crystalline structures of ice. The most common form of ice or snow is based on the hexagonal lattice structure (I_h). The other ice structural phases are labeled nondescriptly in terms of Roman numerals, except for the slightly less metastable cubic structural phase (I_c).

At everyday atmospheric pressure of 1 atm, water is the only known non-metallic substance that exhibits the property that its crystalline phase has a smaller density than the corresponding liquid phase. In particular, the density of solid ice is 8% less than liquid water. This peculiar property has enabled the observed phenomenon that ice floats above liquid water. Microscopically, the smaller density of ice when compared with that of liquid water is due to the fact that hydrogen bonding between the water molecules are less efficient in a crystalline state in which the system is also subjected to additional structural constraints (hexagonal arrangement). This peculiar property of water has immense ramification on the Earth's climate as well as its ability to support life forms as we know it [3].

Once we increase the temperature, our common intuition dictates that ice would melt into liquid water. Due to the strong hydrogen-bonding between water molecules, liquid water has the highest surface tension of 72.8 mN/m for any non-metallic liquids. Water also has the second highest specific heat capacity, after ammonia, as well as high heat of vaporization.

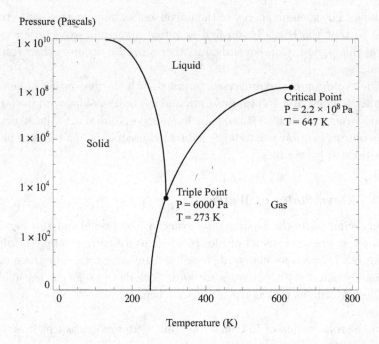

Fig. 2.1   Schematic phase diagram of water. The critical point between the liquid and gas phase represents a second-order phase transition at temperature $T = 647$ K and pressure $P = 2.2 \times 10^8$ Pa. The triple point denotes the condition that all three phases coexist in equilibrium. This only occurs at $T = 273$ K and $P = 6 \times 10^3$ Pa.

Many substances would dissolve in liquid water, which earns it the title of *universal solvent*. When compared with ice, the hexagonal crystalline symmetry has been broken in the high-temperature liquid phase, and the effect of kinetic energy of the molecules come into play. In the liquid phase, we have the delicate interplay between the kinetic energy and the intermolecular potential energy of the system. More specifically, we generally have a competition between the dominant dipole-dipole interaction, orientational-specific hydrogen-bonding, as well as the usual van der Waals interaction.

Since heat has to be imported into the system in order to melt ice into liquid water, we classify this phase transition as *first-order*. The amount of heat needed is known as the *latent heat of fusion*. Generally, there is explicit symmetry breaking in first-order phase transitions. The above example of solid-liquid (water) transition is one such example. We shall elaborate on the details of first-order transitions as well as the specific models of liquid-solid transition in subsequent sections.

Finally, to transition from liquid phase of water to its gaseous phase of water vapor, the system temperature has to be once again increased, thus effectively further drawing energy — *latent heat of vaporization* — into the system and consequently rendering this phase transition first-order as well. If we look at the phase diagram of water in Fig. 2.1, this amounts to crossing the solid transition line between the liquid and gas phases. However, there exists an alternate route to reach the gas phase from liquid phase via the *critical point*. At this critical point the phenomenon of *critical opalescence* occurs. Physically, as one approaches critical opalescence, the size of the gas and liquid regions within the system fluctuates wildly over increasingly large length scales. Once the wavelength of these local inhomogeneous regions becomes comparable to the wavelength of light, scattering occurs and the resulting system goes from transparent to cloudy in appearance. Physical differentiation of the denser liquid phase and the less-dense gas phase occurs due mainly to the effect of gravity. Visually, the boundary meniscus remains the only distinction between the two phases whose origin lies in the different indices of refraction of the two phases. In general, phase transitions such as critical opalescence, which does not involve discontinuous change in the system's free energy induced by the addition of latent heat, is known as *continuous* or *second-order* phase transitions. The main qualitative difference between discontinuous and continuous symmetry is that the latter does not have explicit symmetry breaking, as in the example of the gas and liquid phases of water which only differ in their densities.

Before elaborating further on the general characteristics of first-order and second-order transitions, we consider one more example of phase transition involving discreet degrees of freedom, namely, magnetic spins on a lattice.

2.1.3 *Spins and Magnetism*

Unlike the previous examples illustrating phase transitions in the context of the evolution of our Universe and of the different states of water, the example of spins and magnetism is an example of phase transitions occurring in a discrete system set on a lattice. Whereas the aforementioned examples of the Universe and water are based on systems with a large number (infinite) of degrees of freedom whose critical properties are described by phenomenological models constructed in terms of symmetry principles, we can construct exact models for the system of spins along a one-dimensional, open or closed, lattice as well as that on a two-dimensional square lattice [4, 5].

In general, the construction of these exact models assumes a short-range interaction between spins that is of the nearest-neighbor type. Although sounding extremely limiting at first, these exact models based on magnetic spins on a lattice turn out to have quite a wide range of applications. Examples include adsorption on a linear polymer, on a protein chain, or the elastic properties of fibrous proteins, to name a few [5]. Given the simplicity in the setup of these systems, the exact models that describe them afford us to focus much more clearly on the intrinsic nature of critical phenomena without complicated physical and mathematical details as well as demonstrate their universality that depends mainly on the dimensionality of space and the dimensionality of the order parameters for these systems.

We first consider the conceptually simplest case of having N spins situated on a one-dimensional lattice. Assuming only nearest-neighbor interaction, we can write down its Hamiltonian (energy) of this model, known as the *Ising model*, as follows:

$$\mathcal{H} = -\sum_{\{i,j\}} J_{ij}\sigma_i\sigma_j \tag{2.1}$$

where the set of spins can only take on values of $\{\sigma_i\} = \pm 1$ depending on whether they point "up" (1) or "down" (-1). The notation of $\{i,j\}$ underneath the summation sign denotes summation over only nearest-neighbor spin pairs of i and j. The interaction parameter $\{J_{ij}\}$ indicates the nature of interaction between spins such that when $J_{ij} > 0$, the interaction is ferromagnetic which awards neighboring spins that align along the same direction; when $J_{ij} < 0$, the interaction is antiferromagnetic which awards neighboring spins that align along opposite directions; when $J_{ij} = 0$, neighboring spins do not interact at all.

We note that the Ising Hamiltonian is in fact quite general, as we have not specified the geometry of the lattice upon which the spins reside. The exact solution was first derived by Ernst Ising in 1925, while the exact solution in two dimensions was obtained by Lars Onsager in 1944. Mathematically, with the spin variables taking only one of the two possible values, $\sigma = \{\pm 1\}$, the Ising model possesses a global $\mathbb{Z}_2$-symmetry. Thus, the Ising model is a classical example of a model with a global discrete symmetry. Generalization to $\mathbb{Z}_N$-symmetry can be done in a straightforward manner by letting the spin variables σ to take on N discreet values. An example of a class of models that respect this global $\mathbb{Z}_N$-symmetry includes the N-state Potts model. Its Hamiltonian of the Potts model, given by

$$\mathcal{H}_{\text{Potts}} = -J\sum_{\langle l,l'\rangle}\left(N\delta_{\sigma_l,\sigma_{l'}} - 1\right) \tag{2.2}$$

associates an energy value when neighboring spins are of the same state and a different value when neighboring spins are of different states. The Ising model is thus simply a two-state ($N = 2$) Potts model. A one-state ($N = 1$) Potts model in turn describes the phenomenon of percolation [6].

A closed related class of models to the discrete symmetry-based Ising models and Potts models is the n-vector model , which can be defined as having unit-length, n-dimensional spins s_l at each lattice site, such that $|s_l \cdot s_{l'}| = \delta_{ll'}$. The corresponding Hamiltonian for the n-vector model is as follows:

$$\mathcal{H}_{n\text{-vector}} = -J \sum_{\langle l, l' \rangle} s_l \cdot s_{l'} \qquad (2.3)$$

In contrast to the Ising and Potts model where we have both discrete space symmetry as well as discrete internal symmetry of the $\mathbb{Z}_N$-kind, we have in the n-vector model the case with discrete space symmetry and continuous internal rotation symmetry of $O(n)$. When $n = 0$, we have the model for self-avoiding random walks. When $n = 1$, we retrieve the limit of Ising-model. When $n = 2$, we have the XY-model. When $n = 3$, we have the well-known Heisenberg model that describes the physics of *ferromagnetism* in statistical mechanics.

We remark that the $n = 0$ case may be somewhat counterintuitive at a first glance. However, the mapping between the problem of random walks that are self-avoiding (non-intersecting) and the $n = 0$-limit of the $O(n)$-model will be made clear in Chapter 8, where we consider the statistical properties of topologically-constrained polymers. For now, we simply point out that the mathematical problem of self avoiding random walks is quite useful in modeling the configuration space of chained polymers. In particular, we can formulate a field-theory based on the $O(n)$-model in the limit of $n \to 0$ to describe the critical behavior of polymers. We will consider this mapping further in Chapter 8 in the context of closed-loop polymers in theta solution as an illustration of a renormalization group calculations of varying critical exponents.

2.2 Modern Classification of Phase Transitions

The original Ehrenfest classification scheme considers the non-analyticity property in the free energy of systems of interest. In particular, first-order transitions are characterized by a discontinuity in the first-derivative of the free energy with respect to a thermodynamic variable. In the transitions between the various phases of water, for example, there exists discontinuous changes in the density of the system.

On the other hand, second-order phase transitions are characterized by a discontinuity in the second derivatives of the free energy with respect to a thermodynamic variable. The example of spin magnetism on a discrete lattice exemplifies this mathematical property in that the magnetization, defined as the first derivative of the free energy density with respect to the applied field strength, increases continuously from zero to a finite value as the temperature of the system is lowered, while the magnetic susceptibility based on the second derivative of the free energy with respect to the applied field diverges. In principle, it is theoretically permissible to have phase transitions classified as third- or higher-order.

Modern classification of phase transitions mainly involves whether a given transition involves latent heat. Obviously, first-order transitions do, while second-order transitions do not. There also exists *infinite-order* phase transitions that are continuous but break no symmetries. Examples include the Kosterlitz-Thouless transition in the two-dimensional XY-model as well as a few *quantum phase transitions* [7].

2.3   First-Order Phase Transitions: Solid-Liquid Transition

Unlike critical phenomena characterized by second-order transitions, universality does not exist for first-order transitions. As a result, the details of each system become important. And there generally exists specific models for specific systems. In our aforementioned example of water, we note that the main order parameter characterizing the three states of water is the system density. In particular, the average density $\langle n(\boldsymbol{x}) \rangle$ in the liquid phase is spatially uniform, whereas in the solid phase it is periodic with a Fourier expansion in terms of the reciprocal lattice vectors $\boldsymbol{G}$, as follows:

$$\langle \delta n(\boldsymbol{x}) \rangle = \langle n(\boldsymbol{x}) \rangle - n_0 = \sum_{\boldsymbol{G}} n_{\boldsymbol{G}} e^{i \boldsymbol{G} \cdot \boldsymbol{x}} \qquad (2.4)$$

where n_0 is the average uniform density. A phenomenological free energy that would give rise to the usual observed maximum peak in the structure function S_{nn} of the system is given by

$$F = \int d\boldsymbol{x} d\boldsymbol{x}' \langle \delta n(\boldsymbol{x}) \rangle \chi_0^{-1} \langle \delta n(\boldsymbol{x}') \rangle - w \int d\boldsymbol{x} \langle \delta n(\boldsymbol{x}) \rangle^3 + u \int d\boldsymbol{x} \langle \delta n(\boldsymbol{x}) \rangle^4$$
$$(2.5)$$

where

$$\chi_0^{-1}(x, x') = \left[r + c \left(\nabla^2 + k_0^2 \right)^2 \right] \delta (x - x') \tag{2.6}$$

and $\chi(x, x') = TS_{nn}(x, x')$. As a general feature, the third-order term in the free-energy expansion would generally lead to a first-order phase transitions [4]. The task at hand now is to determine a lattice geometry with a given set of reciprocal lattice vectors G that would minimize the above free energy. This can be achieved by considering the Fourier-transformed free-energy density of Eq. (2.5):

$$f = \sum_{G} \frac{1}{2} r_G \left| n_G \right|^2 - w \sum_{G_1, G_2, G_3} n_{G_1} n_{G_2} n_{G_3} \delta_{G_1 + G_2 + G_3, 0}$$

$$+ u \sum_{G_1, G_2, G_3, G_4} n_{G_1} n_{G_2} n_{G_3} n_{G_4} \delta_{G_1 + G_2 + G_3 + G_4, 0} \tag{2.7}$$

where $r_G = r + c \left(G^2 - k_0^2 \right)^2$. Given the complexity of Eq. (2.7). we see that this task is generally quite complicated. Of course the limitation of any model based on the Taylor expansion of the variation in the order-parameter can be valid only when the order parameter itself does not take on too large of a value. An alternative way to circumvent this difficulty is to follow the approach set forth by the methodology of classical density-functional theory, as will be described in greater details in the next Chapter. In the context of solid-liquid phase transition, this approach was first introduced by Ramakrishnan and Yussouff [8], with a properly chosen reference liquid state.

2.4 Second-Order Phase Transitions: Scaling and Universality

Phases transitions of the second-order kind are continuous in nature. Due to their association with the critical points for the systems, second-order phase transitions are known as critical phenomena. The hallmarks of second-order transitions are the experimental observations of *scaling* and *universality*. Scaling properties, of particular importance scalar invariance, have their physical origin from the fact that fluctuations occur at all length scales as the critical point of a system is approached. Therefore, to describe the physics resulting from these fluctuations, the mathematical models must indeed be scale-invariant. Fortuitously, this required scale-invariance also implies that the particular microscopic degrees of freedom in the system may not matter when the systems approach their critical points. This phenomenon is known as universality. Physically a diverse set of systems may look very different microscopically, yet as they approach a certain critical point their critical behaviors may converge, manifesting in the mathematical forms obeyed by the given set of thermodynamic variables characterizing these systems. This fact allows for tremendous simplification in the study of second-order phase transitions when compared with their first-order counterparts.

To elaborate further, in critical phenomena the only important length scale is the correlation length. Knowing how the correlation length scales with temperature, we can determine all other relevant thermodynamic functions. In particular, these thermodynamic functions are scaled with respect to temperature and length in terms of a set of critical exponents, which depend mainly on the dimensionality of the system as well as its symmetry of the underlying order parameters. Physical systems that are characterized by the same set of critical exponents are said to belong to the same universality class. These critical exponents can be computed theoretically and verified experimentally. A powerful theoretical paradigm known as *renormalization group*, first introduced by Kenneth Wilson in the early 1970s, is based on the concepts of scaling and universality [9].

In the following, we provide a survey of the main ideas and deep insights afforded by the development of renormalization group. The method of renormalization group is certifiably one of the most beautiful and deep ideas borne out of the 20th century modern physics.

2.5   Renormalization Group

In this section, we survey one of the most beautiful and deep concepts arising from 20th-century modern physics. The paradigm of renormalization group was put forth by Kenneth Wilson in 1974 based on the concepts of scaling and universality, for which he won the Nobel Prize in physics. In one line, the mathematical framework of renormalization group details how the properties of a system change with respect to different scales (for example, energy or length).

2.5.1   *Kadanoff Picture: Coarse-Graining of Spin Blocks*

Now we consider the fundamental mechanics of renormalization group. As is conventionally done, the conceptual underlying of renormalization group

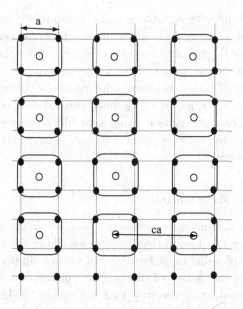

Fig. 2.2   Coarse-graining of Spin Blocks: In the Kadanoff construction, we group four spins in neighboring vertices as a unit and consider their average their overall spin value. The resulting "spin-block" constitute a new fundamental degree of freedom after the averaging (coarse-graining) process. Assuming an initial lattice spacing of a between neighboring spins, the rescaled spin blocks have a rescaled spacing of ca, which is increased by a factor of c.

can be best introduced via the example of block spins. To do so, we return to the system of spins on a lattice in two dimensions. The critical behavior of this system, as a function of temperature, can be determined exactly via the Onsager solution to this two-dimensional Ising model [4, 5] . We will now approach the study of this system in a slightly different angle.

Consider the system as shown in Fig. 2.2. As before, we assume only nearest-neighbor interactions characterized by strength J. The thermo-dynamic properties of this system is also a function of temperature T. Thus, we can write the Hamiltonian of this system, with each spin located at the vertices of the underlying square lattice, as $\mathcal{H} = \mathcal{H}(J, T)$. As usual, we have a competition between the ordered state of spin alignment favored by the J-interaction and the drive towards disorder by the effect of temperature.

However, we can also equivalently describe the same system in terms of 2×2-sized blocks of spins as illustrated in Fig 2.2. Suppose that the lattice spacing between neighboring sites is a. We now assume that the center of mass of each spin block is at a distance of ca from the next-nearest center of mass of its neighboring spin-block units. Presumably, we would average over the total spins inside the block to obtain the same mean value for the effective spin $\sigma' = \langle \sum_i \sigma_i \rangle$, as a function of some new distance scale $y = cx$. In terms of these mean values of total effective spin per block, we can correspondingly ascribe a new effective interaction J' between neighboring spin blocks σ'. Consequently, there must also exists a new temperature scale T' in this new description using spin blocks corresponding to the physics describing the same physical system of spins at lattice sites at the original temperature T.

2.5.2 *General Formulation*

Let us pause for a moment to consider the ramification of the aforementioned two different but equivalent way of describing the same underlying physical system of spins on a lattice. As we are simply using a different mathematical accounting of this original spin system, the true physics should be independent of the way that we choose to describe it. Consequently, the partition function Z_σ of the original description in terms of individual spins on lattice sites should be mathematically equivalent the partition function $Z_{\sigma'}$ in terms of the 2×2-spin blocks, modulo some unimportant overall global rescaling:

$$Z_{\sigma'} \left[J', T' \right] \sim Z_\sigma \left[J, T \right] . \tag{2.8}$$

Any model whose partition function satisfies the mathematical relation in Eq. (2.8) is known to be *renormalizable*. In effect, we are simply performing *coarse graining* on the system. One can imagine repeating the same procedure; this time, we will consider four 2×2-spin blocks and take their average effective spin value σ''. We can analogously scaled the length between these "super"-spin units and define an effective neighboring interaction J'' between these super-spin blocks as well as another new temperature scale T''. In the end, we must have a new relation of the resulting partition function $Z_{\sigma''}$ analogous to Eq. (2.8):

$$Z_{\sigma''} \left[J'', T'' \right] = Z_{\sigma'} \left[J', T' \right] = Z_\sigma \left[J, T \right] . \tag{2.9}$$

and so on.

Let us now rewrite Eqs. (2.8) and (2.9) in a more systematic and general manner explicitly in terms of a given Hamiltonian. In particular, we have

$$Z_\sigma = \sum_{\{\sigma_i\}} \exp \left[-\beta \mathcal{H}_{\{\sigma_i\}} \left(\{J_\alpha\} \right) \right], \quad \alpha = 1, 2, \dots \tag{2.10}$$

where we have allowed for the possibility of having more than one interaction parameters in the set $\{J_\alpha\}$. When we perform coarse-graining, we are rescaling the total number of spins (degrees of freedom) and the correlation length as follows:

$$N' = c^{-D}N, \ \xi' = c^{-1}\xi, \tag{2.11}$$

where we have generalized our discussion to the general space dimensionality of D and denoted the scaling factor c in accordance with our earlier notation. In terms of the new spin-block variables σ', we have the following form for the rewritten partition function in terms of these new spin-blocks:

$$Z_{\sigma'} = \exp\left[N'J'_0\right] \sum_{\{\sigma'_j\}} \exp\left[-\beta\mathcal{H}_{\{\sigma'_j\}}\left(\{J'_\alpha\}\right)\right], \ \alpha = 1, 2, \ldots \tag{2.12}$$

where we have allowed for the fact that there may exist an overall rescaling factor in front. Schematically, we can define an operator $\mathcal{R}$ such that it performs the scaling action in the physical system and mathematically maps $\{J_\alpha\} \rightarrow \{J'_\alpha\}$. In terms of a mathematical parameter space of these interaction parameters, we can rewrite the above statement much more compactly as

$$\boldsymbol{J'} = \mathcal{R}\boldsymbol{J} \tag{2.13}$$

Therefore, Eq. (2.9) can be equivalently represented by the following sequence of operations:

$$\boldsymbol{J}^n = \mathcal{R}\boldsymbol{J}^{(n-1)} = \mathcal{R}^2\boldsymbol{J}^{(n-2)} = \ldots = \mathcal{R}^n\boldsymbol{J}^{(0)}. \tag{2.14}$$

The set of interaction parameters $\{\boldsymbol{J}\}$ are often identified as *running coupling constants* and the change in their values is said to constitute mathematically the *renormalization-group flow* in the parameter space J, as specified by the set of β-functions given by $\mathcal{R}\boldsymbol{J} = \beta\left(\{J_\alpha\}\right)$.

Correspondingly, we have the successive transformation on the correlation length of the system:

$$\xi^{(n)} = c^{-n}\xi^{(0)}. \tag{2.15}$$

Similarly, the free-energy density has the following scaling relation:

$$f^{(n)} = c^{nD}f^{(0)}. \tag{2.16}$$

The *fixed points* $\{J^*\}$ of a given (above) set of transformations $\mathcal{R}$ is defined as

$$J^* = \mathcal{R}J^*. \tag{2.17}$$

In terms of the correlation length ξ, the above relation implies that at fixed point J^*, the correlation length can take on exactly two values of $\xi(J^*) = \{0, \infty\}$. While the physically uninteresting limit of zero correlation-length implies a non-interacting system, the limit of infinite correlation length signals precisely the onset of a phase transition. Thus, in the language of renormalization group, physical information regarding the critical behavior of a system can be extracted from an analysis of the existing fixed points of the underlying group flow.

To further examine renormalization group flows in the vicinity of various critical points, we consider slight variation in the parameter vector J as follows:

$$J = J^* + j, \tag{2.18}$$

$$J' = J^* + j'. \tag{2.19}$$

Since

$$J' = \mathcal{R}(J^* + j) = J^* + \mathcal{R}j = J^* + j', \tag{2.20}$$

$$j' = \mathcal{R}j \tag{2.21}$$

we can then linearize the transformation $\mathcal{R}$ in the vicinity of J^* due to the smallness of variation j in order to obtain

$$j' = \mathcal{L}_{\mathcal{R}}j, \tag{2.22}$$

Assuming an eigen-expansion of the linear operator $\mathcal{L}_{\mathcal{R}}$ of the form

$$\mathcal{L}_{\mathcal{R}}\phi = \sum_k \lambda_k \phi_k, \tag{2.23}$$

we can analogously expand j and j' in terms of the same basis $\{\phi_k\}$

$$j = \sum_k u_k \phi_k, \tag{2.24}$$

$$j' = \sum_k u'_k \phi_k, \tag{2.25}$$

such that Eq. (2.22) can be re-expressed as

$$u'_k = \lambda_k u_k .$$ (2.26)

For n iterations, we simply obtain

$$u_k^{(n)} = \lambda_k^n u_k^{(0)} .$$ (2.27)

Now we can classify three possibilities for the behavior of the corresponding renormalization group flow in the vicinity of a given fixed point J^* based on the magnitude of the eigenvalue $\{\lambda_k\}$: if $\lambda_k < 1$, the corresponding interaction parameter or coupling constant J_k is known as *irrelevant*, since after each successive iteration of a scale transformation, the interaction parameter would only get smaller in magnitude and thus become insignificant as one flows away from the fixed point; if $\lambda_k > 1$, the corresponding coupling J_k is deemed *relevant* since its magnitude would increase as one flows away from the fixed point; the case of $\lambda_k = 1$ is known as *marginal* as the magnitude of the corresponding coupling parameter neither decays nor grows as one leaves the critical point. In this last case, in order to determine its true effect on the overall renormalization group flow, we need to further examine the nonlinear contributions of the transformation.

2.5.3 *Critical Exponents*

We now illustrate how the above formulation can be used to determine the critical exponents of the system. For example, let us consider the critical exponent ν that defines the scaling of the correlation length with respect to the reduced temperature $t \equiv \frac{T-T_c}{T_c}$, via the following:

$$\xi \sim t^{-\nu}$$ (2.28)

In light of the scaling relation in Eq. (2.27), we now adapt it to the specific case of correlation length:

$$\xi(u_1, u_2, \ldots) = c^n \xi(\lambda_1^n u_1, \lambda_2^n u_2, \ldots)$$ (2.29)

Assuming that $u_1 \equiv t$, we obtain

$$u_1^{-\nu} = c^n (\lambda_1^n u_1)^{-\nu} ,$$ (2.30)

$$\nu = \frac{\log c}{\log \lambda_1}$$ (2.31)

To see that the critical exponent ν does not explicitly depend on the scaling factor c, we note that if the linearized $\mathcal{R}$-transformations must satisfy the property of

$$\mathcal{L_R}\left(c_i\right)\mathcal{L_R}\left(c_j\right) = \mathcal{L_R}\left(c_i c_j\right), \qquad (2.32)$$

since a scale transformation by a factor of c_j followed by c_i must be the same as the overall scale transformation by a factor of $c_i c_j$, the eigenvalues of the $\mathcal{L_R}$ must in turn be of the form

$$\lambda_k = c^{\omega_k} \qquad (2.33)$$

Thus, we can ultimately rewrite the result in Eq. (2.28) without a seeming dependence on the scale factor:

$$\nu = \frac{1}{\log \omega_1}. \qquad (2.34)$$

By examining the other scaling relations that define the various other critical exponents, we can similarly evaluate the explicit expressions of the other critical exponents using the apparatus of renormalization group in a rigorous manner [4, 5, 10, 11].

2.5.4   *Origin of Universality Class*

In the previous subsection, we see how the construction of renormalization group provides a rigorous method of computing the various critical exponents. We will now tie up the last loose thread in seeing how the conceptual framework of renormalization group enables a top-down understanding of the phenomenon of universality associated with critical phenomena.

In the language of renormalization group, we see that the critical exponents, whose behaviors encode critical properties of the underlying physical systems, are controlled only by the relevant variables close to various critical points. Physically, the above analysis translates into the fact that the critical behavior of any macroscopic system, in spite of possessing a vast large number of degrees of freedom, is determined by only a handful of these relevant parameters. The remaining details regarding the system particulars do not have significant consequences in determining the critical behavior of these systems. Therefore, we can now see why seemingly divergent macroscopic systems can behave similarly when they approach their respective critical states, as characterized by their identical set of critical exponents. Most of the system details are literally irrelevant when determining their critical behaviors when examined using the conceptual framework of renormalization group.

Systems obeying the set of critical exponents are said to belong to the same *universality class*. One of the great simplification conferred by the existence of universality in second-order transitions is our necessity to only examine the various universality classes without regard to the microscopic details of widely different physical systems. One can conceptualize universality as somehow a consequence of diverging correlation length, which minimizes and erases the importance of specific microscopic interactions between the systems' constituents. This *loss of details* when one approaches critical points can also been seen from the fact that the renormalization operator $\mathcal{R}$ simply does not have an inverse. In other words, once we coarsegrain the system, we cannot reverse the procedure. Thus, the framework of *renormalization group* actually possesses the mathematical structure of a *semi-group* rather than the misnomer found in its name. We will discuss the finer mathematical points regarding renormalization group in the next section.

For now, we simply point out that due to universality, second-order transitions are much better understood than their first-order counterparts in which the simplification and unification afforded by universality does not exist; we have to construct a particular model for the critical behavior of each physical system undergoing first-order transitions.

2.5.5 Wilsonian Picture: Momentum-Space Renormalization Group

So far in our discussion, we have considered models formulated in the real space. Most often when we work with field theories, it is much more advantageous to describe our systems in the momentum space instead. We would expand our degrees of freedom, specifically the field variables, in Fourier expansion. The procèdure of coarse-graining becomes *integrating out momentum shells* in the momentum space. In particular, coarse-graining small length scales is equivalent to integrating out high momentum modes beyond a certain cutoff Λ in momentum space.

The framework of renormalization group in momentum space was presented by Kenneth Wilson in 1974. The momentum-space framework is the natural setting for the theories of three of the four fundamental forces in particle physics, namely, quantum electrodynamics (QED), the Weinberg-Salam model for the electroweak force, as well as quantum

chromodynamics (QCD). There exists well developed perturbation proce-
dures to compute various renormalized coupling constants and field correlations in terms of the corresponding β-functions as well as the Callan-
Symanzik equation [10, 11].

In Chapter 8, we will present an explicit calculation using the
momentum-space renormalization group for the example of topologically-
constrained polymers in theta solution. In particular, we will determine the
various critical exponents for this system and thus the universality class to
which it belongs. We will therein demonstrate in full force the power and
elegance of renormalization-group techniques. In the meantime, we will conclude this section by further examining the mathematical underpinnings of
critical phenomena and the renormalization-group framework, the general
analysis of fixed points in mathematical transformations.

2.6   Mathematical Miscellanies: Semi-Group Structure and Fixed-Point Theorems

2.6.1   *Semi-groups*

As previously mentioned, we inherently throw out information about the
physical system during the operation of scale transformation in the context of coarse-graining. Mathematically, there does not technically exist
an inverse operation that would reverse a renormalization-group flow in a
given parameter space. Consequently, the name *renormalization group* is
technically a misnomer, as the lack of an inverse technically endows the
operation the structure of a *semi-group*. By definition, a semi-group is a
mathematical structure given by

Definition 2.1. A semi-group $(G, *)$ with identity is a set G with a binary
operation $*$ that satisfies the four axioms below:

Identity: There exists an element e in G, called the identity, such that for
all a in G, $e * a = a * e = a$.

Closure: For all elements a and b in G, $a * b \in G$.

Associativity: For all elements a, b, and c in G, $(a * b) * c = a * (b * c)$.

We note that a semi-group with identity as defined above is also sometimes referred to as a *monoid*, while the term semi-group can be associated

with a more general mathematical structure that does not necessarily include an identity element. The important distinction here is that the lack of an inverse makes the semi-group with identity a more general structure than a *group* structure familiar in theoretical physics.

2.6.2 *Miscellany on Fixed-Points*

In the context of momentum-space renormalization group, there are two possible classifications of behavior for the running coupling constants or other interaction parameters in the parameter space based on an analysis of their corresponding β-functions. When the interaction parameter tends to a fixed-point value at low energy scale, we refer to it as the *infrared fixed point*. In real space, low-energy scale corresponds to large length scale. On the other hand, when a fixed point is approached by the interaction parameter in high-energy scale, we refer to it as the *ultraviolet fixed point* that coincides with small length scale in real space. As β-function details the scale dependence of the interaction parameters by definition, determining the locations of these fixed points involving finding zeroes of the corresponding β-functions often order-by-order in perturbation. We will perform this computational exercise for the case of topologically constrained polymers in theta solution and demonstrate the results in Chapter 8.

To conclude this subsection, we briefly mention that there are a few mathematical results regarding fixed-points analysis of general classes of transformations. For the most part, mathematicians are concerned about the conditions under which fixed points would exist for given transformations, i.e. $F(x) = x$. For example, in finite-dimensional spaces, the *Brouwer fixed point theorem* indicates that every continuous function of a n-dimensional closed unit ball D^n to itself has at least one fixed point. The *Lefschetz fixed-point theorem* generalizes to the case of continuous mapping from a compact topological space to itself. Furthermore, for cases of infinite-dimensional spaces, the *Schauder fixed point theorem* applies to continuous mapping from a nonempty convex subset of a Banach space to itself.

2.7 Conclusion

In this chapter, we have provided a survey of several important concepts on phase transitions. Many phase transitions of interest can be categorized either as first-order (discontinuous) or second-order (continuous). First-order phase transitions involve a discontinuous change in its system free energy in the form of latent heat, and they occur between two phases that possess different symetries. In general, less is known about them as we need to provide a particular model for each specific physical system undergoing first-order transitions under study. On the other hand, second phase transitions occur at physical systems whose two phases do not involve explicit symmetry breaking, and they are characterized by changes in system free energy that are continuous. Due to the phenomenon of universality, second-order phase transitions are much better understood when compared with their first-order counterparts. The paradigm of renormalization group provides an important unifying framework from which to understand concepts of scaling and universality that are signatures of second-order transitions.

In the language of renormalization group, universality arises from the fact that most microscopic details of the physical system are affiliated with irrelevant variables in the parameter space of renormalization group flows. These variables decay in magnitude as we "flow" away from a fixed point in the flow. In contrast, relevant variables become more and more important as we approach these group-flow fixed points. The fixed points of parameter flows correspond physically to the critical points of phase transitions, which is characterized by the property of scale invariance due to diverging correlation lengths.

Renormalization group proves to be a valuable method in calculating various critical exponents and correlation functions in both real and momentum space. In Chapter 8, we will present an explicit calculation to determine the universality class for topologically constrained polymers in theta solution. There we will utilize the momentum-space renormalization group to compute the various critical exponents based on a field theory.

References

[1] D. R. Chialvo, P. Balenzuela, and D. Fraiman, *Proceedings of BIOCOMP2007 — Collective Dynamics: Topics on Competition and Coorperation in the Biosciences*, Vietri sul Mare, Italy (2007).

[2] D. R. Chialvo, *New Ideas in Psychology* (2008).

[3] J. Hoffman, T. Tin, and G. Ochoa, *Climate: The Force that Shapes Our World and the Future of Life on Earth*, Rodale Books, New York, 2005.

[4] P. M. Chaikin and T. C. Lubensky, *Principles of Condensed-Matter Physics*, Cambridge University Press, New York, 2000.

[5] R. K. Pathria, *Statistical Mechanics*, 2nd ed., Butterworth-Heinemann, 2000.

[6] B. Bollobas and O. Riordan, *Percolation*, Cambridge University Press, Cambridge, 2006; M. Sahimi, *Applications of Percolation Theory*, CRC, New York, 1994.

[7] S. Sachdev, *Quantum Phase Transitions*, Cambridge University Press, Cambridge, 2001; W. G. Unruh and R. Schutzhold, *Quantum Analogues: From Phase Transitions to Black Holes and Cosmology*, Springer, New York, 2007.

[8] T. V. Ramakrishnan and M. Yussouff, *Phys. Rev. B* **19**, 2775, 1979.

[9] K. R. Wilson and J. Kogut, *Phys. Reports* **12C**, 75 (1974).

[10] M. E. Peskin and D. V. Schroeder, *An Introduction to Quantum Field Theory*, Addison-Wesley, New York, 1995.

[11] M. Kaku, *Quantum Field Theory, A Modern Introduction*, Oxford University Press, Oxford, 1993.

Chapter 3

Overview of Density-Functional Theory

3.1 Introduction

Soft materials are often subjugated into classes: "colloids, polymers, or emulsions," or "liquid crystals, membranes, or foams." While these systems all share common mechanical and energetic scales, each subfield has developed its own set of theoretical tools, often required and inspired by real-world applications. In this work, we have combined ideas from these subfields to study colloids [1–3], which, in addition to the standard excluded-volume interactions, softly repel each other, either through a surface polymer coating ("fuzzy") [4] or via a screened Coulomb interaction [5]. While free-volume theory provides remarkably accurate accounting of the free energy of packing the hard cores [6], additional pairwise interactions are hard to implement analytically to determine in detail the phase behavior of these systems. It is our goal that a conceptually intuitive framework that, at the same time, readily lends itself to analytically tractable description be proposed and demonstrated as a viable alternative to the orthodox means of modeling these soft colloidal systems; namely, the density functional theory.

3.2 Electronic Density-Functional Theory

Density functional theory is the traditional method in the investigation of the thermodynamic properties of the solid phases of typical soft systems such as colloids. Going back to our discussion on the necessity of approximate methods as well as numerical simulations, we can view density-functional theory in the grand scheme of modeling approaches as an intermediary approach between the conceptual brute force of molecular simulation and the encompassing theoretical elucidation of systems details [15].

While density-functional theory provides computationally a less expensive route to probing collective properties of a system, it does take into account inputs of molecular details of the system as encoded empirically in the density and thereby make it more rigorous than purely phenomenological theories [15–17]. Before presenting our new theoretical formalism of the foam analogy, we shall briefly outline this traditional method in modeling interactions and correlations between particles in condensed-matter systems. Summarily, in the framework of density-functional theory, information is derived for an *inhomogeneous* solid phase from the direct correlation function of the corresponding *homogeneous* fluid phase.

Historically, density-functional theory has its origin in quantum mechanics on the problem of determining the ground state for many-electron systems. Of course, the problem of such determination for these systems would be an example of the quantum version of the many-body problem previously discussed in the Introduction. As is typical in these problems, the complexity lies in the interaction or couplings between the many constituents of the system, which in this context, would include the exchange interaction and correlations between the many electrons in this problem. Instead of relying on the multi-electron wavefunction, labeled by $\Psi_{GS}(r_i)$ with particle index i, as in the fundamental description of the (ground) state of the system found, for instance, in the traditional Hartree-Fock theory, the main utility of density-functional theory is to focus instead on the electronic density, defined by $n_{GS}(r) = |\Psi_{GS}(r_i)|^2$ [18], as the fundamental quantity in the description of the ground state. Due to the remarkable N-fold reduction in the degrees-of-freedom involved, the electronic density $n_{GS}(r)$ is much more conceptually and computationally attractive as an alternative, as well as being much more practical and relevant of a quantity to study when comparing with experimental data given its role in such concepts as chemical potential, response function, structure factor, and bulk modulus.

The validity of the density-functional approach was firmly established by the *Hohenberg-Kohn Theorems*. Their first theorem shows that all ground-state properties of a system are determined by the electronic density, as there essentially exists a one-to-one mapping between the ground-state wavefunction and the corresponding electronic density: $\Psi_{GS}(r_i) \leftrightarrow n_{GS}(r)$; while their second theorem states that a variational principle can be formulated such that the aforementioned ground-state electronic density minimizes the ground-state energy functional. To illustrate these concepts more concretely, we consider the Hamiltonian of a many-electron system,

$$\hat{H} = -\frac{1}{2}\sum_{i=1}^{N}\boldsymbol{\nabla}_i^2 + \sum_{i<j}\hat{U}(\boldsymbol{r}_i,\boldsymbol{r}_j) + \sum_{i=1}^{N}\hat{V}_{ext}(\boldsymbol{r}_i) \tag{3.1}$$

$$\equiv \hat{T} + \hat{U} + \hat{V}, \tag{3.2}$$

where the kinetic energy- and interaction energy-functionals together constitute the *universal functional* found in the original Hohenberg-Kohn Theorems, as the expression is valid for all systems with interacting electrons.
Different physical systems, in turn, are then characterized by the different
one-body external potential $\hat{V}(\boldsymbol{r}_i)$. Thus, the Hohenberg-Kohn Theorems
demonstrates in essence that there exists a one-to-one correspondence between the external potential $\hat{V}(\boldsymbol{r}_i)$ and a ground-state electronic density
$n_{GS}(\boldsymbol{r})$.

Given the correspondence mapping of $\Psi_{GS}(\boldsymbol{r}_i) \leftrightarrow n_{GS}(\boldsymbol{r})$, the total energy functional associated with Eq. (3.2) can thus be expressed equivalently
in terms of the electronic density, $E[\Psi_{GS}(\boldsymbol{r}_i)] = E[n_{GS}(\boldsymbol{r})]$, and it takes
the form:

$$E[n] = T[n] + U[n] + \int d^3r V(\boldsymbol{r})n(\boldsymbol{r}), \tag{3.3}$$

where

$$E[n] \equiv \left\langle \Psi_{GS}[n_{GS}] \left| \hat{H} \right| \Psi_{GS}[n_{GS}] \right\rangle, \tag{3.4}$$

$$T[n] \equiv \left\langle \Psi_{GS}[n_{GS}] \left| \hat{T} \right| \Psi_{GS}[n_{GS}] \right\rangle, \tag{3.5}$$

$$U[n] \equiv \left\langle \Psi_{GS}[n_{GS}] \left| \hat{U} \right| \Psi_{GS}[n_{GS}] \right\rangle, \tag{3.6}$$

are the expectation values of the total-, kinetic-, and Coloumb-energy func-
tionals in terms of the electronic density. To estimate the universal functional consisting of the kinetic and interaction contributions, we can now
apply the *Kohn-Sham approach* [14] and translate the description of our
many-electron system from one where we treat interacting electrons in a
real potential to a corresponding theoretical alternative where we have non-
interacting fictitious counterparts in an effective potential, as follows:

$$T[n] = \sum_i \frac{-h^2}{2m} \int d^3\boldsymbol{r} \, \Psi_i^* \boldsymbol{\nabla}^2 \Psi_i \,, \tag{3.7}$$

$$U[n] = \frac{e^2}{2} \iint \frac{n(\boldsymbol{r})n(\boldsymbol{r}')}{|\boldsymbol{r} - \boldsymbol{r}'|} \, d^3\boldsymbol{r} \, d^3\boldsymbol{r}' + V_{xc}[n] + V[n] \,, \tag{3.8}$$

where the first term in Eq. (3.8) represents the tradition Hartree term representing the electron-electron Coulomb repulsion, and $E_{xc}[n]$ denotes the quantum exchange correlation potential between the electrons beyond the Hartree approximation. To determine the ground state, we must now minimize the energy functional in Eq. (3.3) with respect to the constraint of total fixed number of electrons in the system, as follows:

$$\frac{\delta}{\delta n(\boldsymbol{r})} \left(E\left[n(\boldsymbol{r})\right] - \mu \int n(\boldsymbol{r}) d^3\boldsymbol{r} \right) = 0 \,, \tag{3.9}$$

where the Lagrange multiplier μ corresponds to the chemical potential. Upon the Kohn-Sham approximate expressions for the kinetic and Hartree energy-functionals in Eqs. (3.7) and (3.8), we arrive at the Kohn-Sham equations, as follows:

$$\left[-\frac{1}{2} \boldsymbol{\nabla}^2 + U\left(n_{GS}; \boldsymbol{r}\right) + V\left(n_{GS}; \boldsymbol{r}\right) + V_{xc}\left(n_{GS}; \boldsymbol{r}\right) \right] \Phi_i\left(\boldsymbol{r}\right) = \epsilon_i \Phi_i\left(\boldsymbol{r}\right) \,, \tag{3.10}$$

where $\mu = \epsilon_N$ and all other ϵ_i are simply Lagrange parameters without physical meaning for the corresponding trial orbitals $\Phi_i(\boldsymbol{r})$. Equation (3.10) can be solved in a self-consistent way if we know the exact expression for the exchange-correlation potential V_{xc}. To do so, we make one additional assumption in that the electron density is slowly varying, and we can apply the local density approximation (LDA) to estimate the exchange energy between electrons based on the equivalent quantity for a uniform electron gas, upon which we can determine the effective potential based on the minimization of the corresponding energy functional.

To summarize thus far, the underlying motivation for DFT is to devise a reasonable approximation that can tackle the difficulty of two main defining features of many-body problems: correlations between the constituents of the system and the necessary complexity that arises for their description. DFT provides a computational recipe that interpolates between constituent structure and the resulting collective properties. Not surprisingly, the commonality shared by these many-electron quantum systems in hard-condensed matter physics and the many-particle systems in soft-condensed matter physics, both under the umbrella of many-body problems, makes the philosophy and principles behind DFT much more widely applicable to other physical systems outside of its origin in quantum mechanics. The breadth of applications involving DFT is demonstrated by its use in the study of interfacial properties [19], solvation [20], as well as phase properties of liquid crystals [21], of polymers, and of colloids [22], just to name a few.

3.3 Classical Density-Functional Theory

Much like its electronic counterpart, classical DFT works for a given open system with fixed temperature T, total volume V, and chemical potentials μ_i for each species, due to the underlying one-to-one correspondence between the equilibrium density of molecules, or their spatial distribution thereof, and the external potential for each species. Comparing with its quantum electronic counterpart, the density of electrons has been replaced now with the density of molecules or of other relevant coarse-grained elements of interest. Thus, drawing a parallel with the electronic DFT, classical DFT again makes use of the approximation based on a homogenous density function, which describes the liquid state in the context of soft matters, in order to derive information on inhomogeneous systems that are subject to an external field or for anisotropic fluids such as liquid crystals [15, 20].

As mentioned in the introduction, the free energy in statistical mechanics captures in theory all information pertaining to its equilibrium thermodynamics. Traditionally, the Helmholtz free energy is divided into an ideal part and an excess contribution. The ideal part consists of the contributions attributable to an ideal gas where all intermolecular interactions vanish; the excess contribution then takes into account all such neglected interactions between molecular components not included in the ideal part of the free energy:

$$F = F^{id} + F^{ex} \tag{3.11}$$

For all statistical systems, the ideal part of the free energy is known by definition and takes the following form:

$$F^{id} = k_B T \sum_i \int d\boldsymbol{R}\, \rho_i(\boldsymbol{R}) \log\left[\frac{\rho_i(\boldsymbol{R})}{e}\right] + \sum_i \int d\boldsymbol{R}\, \rho_i(\boldsymbol{R}) V_i(\boldsymbol{R}), \tag{3.12}$$

where the first term consists of the contribution from entropy and the second term takes into account the energy associated with a one-body external potential $V_i(\boldsymbol{R})$ acting on particle i, or the internal molecular energy states associated with each particle i with contributions from their stretching, bending, and vibrational modes. We note that while the expression in Eq. (3.12) can be readily written down analytically, its evaluation would most often require numerical methods that are nevertheless not too computationally intensive.

The excess free energy, F^{ex}, on the other hand, is not always known beforehand for most general systems. Various approaches have been developed specifically for different types of soft systems of interest [15]. The unifying thread underlying the different approaches is the assumption that the excess Helmholtz free energy only depends on the densities of the various components:

$$F^{ex} = F^{ex}\left[\rho_i(\boldsymbol{r})\right] . \tag{3.13}$$

Again, we observe that there is a great reduction of the total number of degrees of freedom needed, from $3N$ to N-components, in the assumption stated above in Eq. (3.13). The contributions to this excess free energy include van der Waals attraction, electrostatic interaction, as well as other interactions of chemical nature [15]. In such system as polymers in solution, the excess free energy also includes such effects as chain connectivity. In the study of phase transitions of colloidal systems, with which this work is mainly concerned, the excess free energy denotes the inhomogeneity found in the lattice phase, while the ideal free energy refers to the homogenous and isotropic state of liquid. We remark that, in practice, different characterization of the excess free energy based on the particular systems and whose properties under study would correspond to the *different* density-functional theories. One famous example is that of Ramakrishnan and Yussouff (RY) on the theory of freezing, which involves the expansion of the excess Helmholtz free energy about the density of a uniform liquid truncated to the second-order [23]. While this truncation seems arbitrary without prior justification, other versions of DFT provides a more rigorous basis by making use of a uniform reference system chosen at an *effective* density based on various criteria related to the system properties of interest. One such example is the weighted density approximation (WDA) [24]. In WDA, the approximate expression for the excess free energy takes the form:

$$F_{WDA}^{ex}[\rho] = \int d\boldsymbol{r}\, \rho(r) f_0^{ex}\left((\bar{\rho}(\boldsymbol{r}))\right), \tag{3.14}$$

where f_0^{ex} denotes the excess free energy per particle of a uniform state (liquid) and $\bar{\rho}(\boldsymbol{r})$ is the *weighted density*, defined with respect to a weighting function, $w(\boldsymbol{r} - \boldsymbol{r}')$,

$$\bar{\rho}(\boldsymbol{r}) = \int d\boldsymbol{r}'\, \rho(\boldsymbol{r}')\, w(\boldsymbol{r} - \boldsymbol{r}'), \tag{3.15}$$

which satisfies the following normalization condition:

$$\int d\boldsymbol{r}'\, w(\boldsymbol{r} - \boldsymbol{r}') = 1. \tag{3.16}$$

To determine this weighting function uniquely, we rely on the relation between the functional derivative of the free energy and the nth-order direct correlation function $c_0^{(n)}(\boldsymbol{r}, \boldsymbol{r}')$ in the following limit:

$$\lim_{\rho_s \to \rho} -\frac{1}{k_B T} \frac{\delta^n F^{ex}}{\delta \rho_1(\boldsymbol{r}) \ldots \delta \rho_n(\boldsymbol{r})} = c_0^{(n)}(\boldsymbol{r}_1, \ldots, \boldsymbol{r}_n),. \tag{3.17}$$

The relation in Eq. (3.17) becomes exact when the inhomogenous density ρ_s of the solid phase approaches that of the actual liquid $\rho_s = \rho$. When we demand that the relation in Eq. (3.17) is *exact* for $n = 2$, we can uniquely determine the weighting function $w(\boldsymbol{r} - \boldsymbol{r}')$. To do so, we note that upon substituting Eq. (3.14) into Eq. (3.17), we can obtain the following differential equation, in Fourier space, satisfied by the weighting function.

$$k_B T c_0^{(2)}(k) = 2f_0^{'ex} w(k) + \rho_0 f_0^{''ex} w^2(k) + 2\rho_0 f_0^{'ex}(\rho_0) w'(k) w(k), \tag{3.18}$$

Now with the weighting function $w(\boldsymbol{r} - \boldsymbol{r}')$ uniquely determined, we can write down the complete excess Helmholtz free energy in Eq. (3.14) with known $\rho(\boldsymbol{r})$ and $f_0^{ex}[\bar{\rho}]$ for the uniformed state evaluated at $\bar{\rho}(\boldsymbol{r})$ via Eq. (3.15). While WDA is conceptually a more-satisfying framework than Ramakrishnan and Yussouff's version of the density-functional theory, it is computationally very expensive. As a result, a modified version of WDA, the modified weighted density approximation (MWDA) [7, 8], proves to be much more useful numerically in practice.

As mentioned on several occasions prior, this work is mainly concerned with a new paradigm of constructing the free energy for the solid phases of colloids that is readily amenable to analytical solutions and that also provides, at the same time, an intuitive way of relating the microscopic composition and details of the system to the various macroscopic thermodynamic properties. In the language of MWDA, the density plays a central role in the formalism. It is the order parameter that characterizes the different phases of the system. And since most often different phases are characterized by their different geometric properties, the density would naturally have to reflect this fact. For the solid phase of molecular lattices, the density is in general represented by a sum of normalized Gaussians centered about their lattice sites R:

$$\rho_s(r) = (a/\pi)^{3/2} \sum_R \exp(-\alpha|r - R|^2). \qquad (3.19)$$

The Fourier components then become

$$\rho_G = \rho_s \exp(-|G|^2/4\alpha), \qquad (3.20)$$

where G denotes the reciprocal lattice vectors for the solid phase under study. Here in the formalism of classical DFT, the geometry and the density of the solid phase are input parameters for the reciprocal lattice vectors G. The parameter α is a measure of non-uniformity within the structure of the system and continuously varies between the limit of uniform liquid ($\alpha = 0$) and of particles frozen in their lattice sites ($\alpha \to \infty$). The method of MWDA then essentially translates the above statement into the satisfaction of the following constraint [8]:

$$\hat{\rho} = \rho_s \left[1 - \frac{1}{2\beta f_0^{'ex}(\hat{\rho})} \sum_G \exp(-|G|^2/2\alpha)C(G;\hat{\rho}) \right], \qquad (3.21)$$

where $C(G)$ is the Fourier transform of the direct correlation function evaluated at the respective G of the reciprocal lattice for a given effective density $\hat{\rho}$, and $\beta = k_B T$ as usual. As before, the quantity $f_0^{ex}(\rho)$ is simply the excess free-energy of the inhomogeneous solid equal to that of the liquid evaluated at some predetermined effective density:

$$f_0^{ex}(\rho) = \frac{F^{ex}}{N}. \qquad (3.22)$$

Experimentally, the direct correlation function can be in general determined from the static structure factor using scattering techniques. In the end, the direct correlation function can be found using the following relation:

$$C(q) = \frac{1}{\rho[1 - \rho S(q)]}.$$ (3.23)

To summarize, in the formalism of MWDA, which is a widely used example of a general class of theories known as DFT, the lattice types and geometric parameters of the solid phase along with the density are input parameters entered by hand. And other parameters encoding relevant experimental conditions enter the formalism indirectly via such quantities as the direct correlation function, which is in turn based on the directly measurable static structure factor [8].

3.4   Conclusion

In this chapter, we see that DFT provides a computational shortcut when compared with brute-force numerical simulations. It also lends rigor to the bottom-up approach in a self-consistent theoretical framework by incorporating some molecular details of the underlying system. However, DFT does not provide any inherent insights to understanding the underlying mechanism of phase transitions in the underlying physical systems under study, nor is its implementation particularly simple given the multidimensional integrations of various quantities in the configuration space.

We note that there are other approaches to studying phase transitions in colloidal systems. Molecular dynamics simulations based upon brute-force computation of the Newton's force law is another traditional route. While these tools have proven important in the advancement of our current understanding of phase transitions in colloids and other soft systems, they do not provide a robust explanation of the stability of colloidal crystals.

In light of the aforementioned shortcomings regarding DFT and other traditional methods, we propose in the next chapter an alternative framework in understanding the thermodynamic properties of the solid phases observed in commonly found fuzzy and charged colloidal systems. We will demonstrate that geometry would play a central role in this new framework, and the various experimental parameters such as particle density would enter into the framework in a much more direct and natural way.

References

[1] P. Ziherl and R. D. Kamien, *Phys. Rev. Lett.* **85**, 3528 (2000).

[2] P. Ziherl and R. D. Kamien, *J. Phys. Chem. B* **105**, 10147 (2001).

[3] W. Kung, P. Ziherl and R. D. Kamien, *Phys. Rev. E* **65**, 050401(R) (2002).

[4] R. L. Whetten, M. N. Shafigullin, J. T. Khoury, T. G. Schaaff, I. Vezmar, M. M. Alvarez and A. Wilkinson, *Acc. Chem. Res.* **32**, 397 (1999).

[5] E. B. Sirota, H. D. Ou-Yang, S. K. Sinha, P. M. Chaikin, J. D. Axe and Y. Fujii, *Phys. Rev. Lett.* **62**, 1524 (1989).

[6] W. A. Curtin and K. Runge, *Phys. Rev. A* **35**, 4755 (1987).

[7] A. R. Denton and N. W. Ashcroft, *Phys. Rev. A* **39**, 4701 (1989).

[8] G. A. McConnell and A. P. Gast, *Phys. Rev. E* **54**, 5448 (1996).

[9] T. L. Hill, *Statistical Mechanics*, McGraw-Hill, New York, 1956.

[10] J. A. Barker, *Lattice Theory of the Liquid State*, Pergamon Press, Oxford, 1963.

[11] We thank M. O. Robbins for discussions on this point.

[12] V. S. K. Balagurusamy, G. Ungar, V. Percec and G. Johansson, *J. Am. Chem. Soc.*, **119**, 1539 (1997).

[13] T. C. Hales, *Notices Amer. Math. Soc.*, **47**, 440 (2000).

[14] W. Kohn and L. J. Sham. *Phys. Rev.*, **140**, 1133A 1965.

[15] J. Z. Wu, *AlChE J.* **52**, 1169 (2006).

[16] M. Rex, H. H. Wensink, and H. Lowen, *Phys. Rev. E* **76**, 021403 (2007).

[17] J. Kleis, B. L. Lundqvist, D. C. Langreth, and E. Schroder, *Phys. Rev. B* **76**, 100201(R) (2007).

[18] S. Datta, *Electronic Transport in Mesoscopic Systems*, Cambridge Univ. Press, New York, 1995.

[19] P. C. Hiemenz and R. Rajagopalan, *Principles of Colloid and Surface Chemistry*, CRC, New York, 1997.

[20] B. Mennucci and R. Cammi, *Continuum Solvation Models in Chemical Physics: From Theory to Applications*, Wiley Press, New York, 2008; F. Hirata, *Molecular Theory of Solvation*, Springer Publishing, New York, 2003.

[21] P. G. de Gennes and J. Prost, *The Physics of Liquid Crystals*, Oxford University Press, Oxford, 1995; P. J. Collings, *Liquid Crystals: Nature's Delicate Phase of Matter*, Princeton University Press, Princeton, 2001.

[22] W. B. Russel, D. A. Saville, and W. R. Schowalter, *Colloidal Dispersions*, Cambridge University Press, Cambridge, 1992; R. J. Hunter, *Foundations of Colloid Science*, Oxford University Press, Oxford, 2001.

[23] T. V. Ramakrishnan and M. Yussouff, *Phys. Rev. B* **19**, 2775 (1979).

[24] W. A. Curtin and N. W. Ashcroft, *Phys. Rev. A* **32**, 2909 (1985).

[25] P. M. Chaikin and T. C. Lubensky, *Principles of Condensed Matter Physics*, Cambridge University Press, New York, 2000.

[26] W. D. Callister Jr., *Materials Science and Engineering: an Introduction*, John Wiley & Sons, New Jersey, 2003.

Chapter 4

Survey of Solid Geometry and Topology

4.1 Introduction

In this chapter, we introduce our geometric approach to constructing the free energy of the different lattice phases. We will then apply our formalism to the specific context of colloidal systems in the next chapter. Fundamentally, our model provides a one-to-one mapping between the free energy of a given lattice structure to that of the corresponding *dual* mathematical structure of *foams*. This one-to-one mapping is furnished by implementing the procedure of *Wigner Seitz-cell* construction. In the literature, Wigner-Seitz cells are also referred to as *Voronoi cells*. Thus, we contextually refer to our general geometric formalism of constructing solid free energy as the *foam model*.

In what follows, we will explain in detail the salient features of the mapping between colloidal crystalline lattices and foams in our model. Furthermore, we will specialize in the case where the lattice points are occupied by colloidal particles. Specifically, we consider two particular types of colloids: charged colloids and fuzzy colloids. Charged colloidal particles interact via a long-range screened-Coulomb potential screened Coulomb interaction, while fuzzy colloids are made of molecules that possess long, flexible alkyl chains attached to the center of rigid, aromatic rings and interact with a simple hardcore potential dressed with a short-range repulsive potential.

As is conventionally done, physicists model colloidal particles with molecular shape isotropy generally as mathematical spheres. For other systems that possess molecular shape anisotropy, such as liquid crystalline mesogens, we can in turn choose other mathematical objects that appropriately reflect similar geometric properties as the actual molecules as their real-life abstraction; in the case of liquid crystals, we generally model the mesogens as cylinders or rods. Going back to our case of colloidal particles, the typical approach undertaken in studying their interactions is to consider the infinite-curvature limit of treating them as point particles. In our approach of the foam model, we take the other extreme in which we treat the interactions between the spheres as if they are between flat, zero-curvature surfaces.

4.2   Lattice Symmetry Groups

To illustrate the details of our foam model, it would be instructive to start with a simple system. Let us first consider a system that contains only one type of particles. In a perfect crystal, the colloidal particles are suitably arranged in a periodic lattice known as *Bravais lattice*. Consequently, each particle constitutes a repeating structural unit inside this periodic lattice. We may divide the total volume into equal volume cells, each singly occupied with a mass density equivalent to one particle. which we will refer to as the *unit cell*. The unit cell with the smallest possible volume is known in turn as the *primitive unit cell*. Each point on the lattice (where the particles are located) can be expressed as an integral linear combination of independent *primitive translation vectors*. There are different ways of constructing equivalent sets of primitive unit cells for a given lattice. One such way is the construction of the so-called *Wigner-Seitz cells* [16, 25]. They can be obtained by constructing the perpendicular bisector on all primitive translation vectors emanating from a lattice point, as in Fig. 4.1.

Mathematically, lattices are structures that are invariant under the group of lattice translation. The definition of a mathematical *group* is as follows:

Definition 4.1. A group $(G, *)$ is a set G with a binary operation $*$ that satisfies the four axioms below:

Identity: There exists an element e in G, called the identity, such that for all a in G, $e * a = a * e = a$.

Inverse: For each a in G, there exists an element b in G such that $a * b = b * a = e$.

Closure: For all elements a and b in G, $a * b \in G$.

Associativity: For all elements a, b, and c in G, $(a * b) * c = a * (b * c)$.

(a)

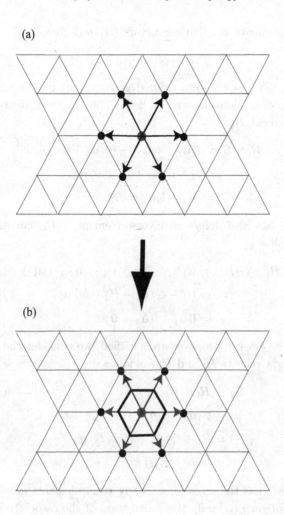

(b)

Fig. 4.1 Construction of Wigner-Seitz cells. We first select a reference lattice point (grey circle). We then construct the perpendicular bisectors on all primitive translation vectors (a). The union of the set of all such perpendicular bisectors result in the corresponding Wigner-Seitz cell enclosing the original reference lattice point (b).

Thus, to see how the operation of translations in a lattice possesses the mathematical structure of a group, we consider for now the case of translations for a two-dimensional lattice. Any lattice point in this two-dimensional lattice can be specified by an integral linear combination of independent primitive translation vectors $\{a_1, a_2\}$ such that

$$R_l = l_1 a_1 + l_2 a_2 \qquad (4.1)$$

where $l = (l_1, l_2)$ is a two-dimensional vector with components $\{l_i\}$. For this operation of translation, we can define the following identity element, 0, for each element R_l:

$$R_l + 0 = (l_1 a_1 + l_2 a_2) + (0, a_1 + 0\, a_2)\,, \qquad (4.2)$$

$$= (l_1 + 0)\, a_1 + (l_2 + 0)\, a_2\,, \qquad (4.3)$$

$$= l_1 a_1 + l_2 a_2 = R_l\,. \qquad (4.4)$$

Moreover, we can also define an inverse element, $-R_l$, corresponding to each R_l, as follows:

$$R_l + (-R_l) = (l_1 a_1 + l_2 a_2) + (-l_1 a_1 - l_2\, a_2)\,, \qquad (4.5)$$

$$= (l_1 - l_1)\, a_1 + (l_2 - l_2)\, a_2\,, \qquad (4.6)$$

$$= 0\, a_1 + 0\, a_2 = 0\,. \qquad (4.7)$$

To see that the closure property are satisfied, we consider the sum of two such translation vectors R_l and $R_{l'}$ as follows:

$$\tilde{R}_l \equiv R_l + R_{l'} \qquad (4.8)$$

$$= (l_1 a_1 + l_2 a_2) + (l'_1 a_1 + l'_2 a_2) \qquad (4.9)$$

$$= (l_1 + l'_1)\, a_1 + (l_2 + l'_2)\, a_2 \qquad (4.10)$$

$$\equiv \tilde{l}_1\, a_1 + \tilde{l}_2\, a_2\,, \qquad (4.11)$$

which obviously has the same form as any other elements in set G. Lastly, it is straightforward to verify the satisfaction of the associativity property via the following relation:

$$\left(\tilde{R}_l + R_l \right) + R_{l'} = \tilde{R}_l + \left(R_l + R_{l'} \right)\,. \qquad (4.12)$$

Thus, we have demonstrated that the operation of translation, characterized by the set of translation vectors $\{R_l\}$, does indeed satisfy all four axioms for the mathematical structure of a group. In general, crystals are also invariant under further sets of symmetry operation known as the *point group*, which comprises rotations, reflections, and inversions upon special symmetry points of the crystalline lattice [16, 25]. Mathematically, a point group is the group of isometry, which is the set of all symmetry operations that leave a point fixed in space. The compatibility requirement between the discreet group of translation in a crystalline lattice and the corresponding isometry group in the d-dimensional space upon which the crystalline lives actually places great constraint on the possible rotational-symmetry elements in the isometry group. According to the *crystallographic restriction theorem*, the rotational symmetries of a crystal in two and three dimensions are indeed limited to two-fold, three-fold, four-fold, and six-fold [26]. When we combine the translational group of a crystalline lattice with the point group associated with each molecule at the lattice points, we obtain the *space group* for the corresponding lattice. In the subsequent sections, we will provide a quick survey of the two-dimensional and three-dimensional space groups and their respective symmetry elements.

4.3 Two-Dimensional Space Groups

In two dimensions, space groups of this dimensionality are also sometimes referred to as *wallpaper groups* or *plane groups*. There are in total 17 wallpaper groups. For each wallpaper group, there generally consists of four elements for the group: *translations*, *rotations*, *reflections*, and *glide reflections*. The last element of glide reflections refers to the operation of a reflection combined with a translation afterwards. The 17 wallpaper groups fall under one of the five basic categories of simple lattices; additional variations are possible depending on how each lattice is decorated with additional elements in each unit cell, which constitutes a repeated unit in the lattice. While the following results are well known, we will give a complete list of them for the sake of completeness [27].

The 17 wallpaper groups in two dimensions can be divided into five general categories based on the geometry of the respective repeating unit cells: oblique lattice, rectangular lattice, centered rectangular lattice, square lattice, and hexagonal lattice, as shown in Fig. 4.2. This fact is not surprising when one considers the possibility of two-dimensional tessellation with only one type of unit cells. In particular, suppose that the two-dimensional lattice under consideration is generated by the translation vector $T = am+bn$, where m and n are basis vectors of the lattice and the unit cell side lengths $|a|$ and $|b|$ are both minimal. The five lattices can then be characterized as follows:

$$oblique \ lattice : |a| < |b| < |a - b| < |a + b| \tag{4.13}$$

$$rectangular \ lattice : |a| < |b| < |a - b| = |a + b| \tag{4.14}$$

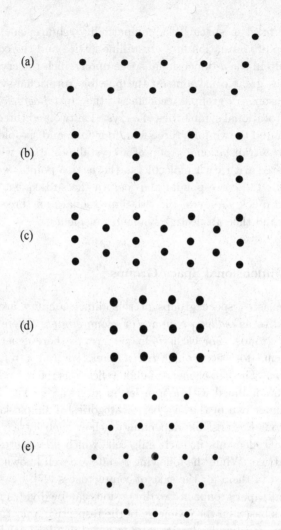

Fig. 4.2   Two-dimensional Lattices.   Oblique lattice (a).   Rectangular lattice (b). Centered rectangular lattice (c). Square lattice (d). Hexagonal lattice (e).

$$centered\ rectangular\ lattice : |a| < |b| = |a - b| < |a + b| \quad (4.15)$$

$$|a| = |b| < |a - b| < |a + b| \quad (4.16)$$

$$square\ lattice : |a| = |b| < |a - b| = |a + b| \quad (4.17)$$

$$hexagonal\ lattice : |a| = |b| = |a - b| < |a + b| \quad (4.18)$$

To denote each of these wallpaper groups, we can employ either the

Hermann-Mauguin notation popular in crystallography or the orbifold notation based on the topological property of these lattices. We shall, in the following, outline both methodologies.

4.3.1 *Hermann-Mauguin Crystallographic Notation*

We first discuss the Hermann-Mauguin crystallographic notations for lattices, which was originally derived for the much more general case of three-dimensional lattices [28]. The notation is denoted by a string of different combinations of four characters composed of either letters or numbers. The logic behind the purpose of this symbolic system is to identify the unit cell of a lattice by all of its symmetry elements. As such, the first letter of the string of four characters distinguishes whether the unit cell is of the primitive (p) or centered (c) type.

A unit cell is of the primitive type when the repeating unit of the lattice has the minimum volume possible and has also lattice points of the highest order of rotational symmetry residing at the vertices of the unit cell. On the other hand, a centered unit cell does not necessarily have the minimum volume for each repeating unit, and the lattice points of the highest order of rotational symmetry reside instead at the center of the unit cell. In the case of the center cell, it is always chosen so that the reflection axis is normal to one or two sides of the centered unit cell.

The next character in the string is an integer (n) denoting exactly what this highest order of rotational symmetry is for such lattice point. In two dimensions, the integer can only be $n = 2, 3, 4, 6$ based on symmetry argument. The next two symbols in the string of four characters refer to possible symmetry elements relative to the the main translation axis of the unit cell. In particular, the third symbol of the Hermann-Mauguin symbol take on the value depending on whether there exists a mirror plane (m), glide plane (g) or none (1). The last symbol in the string of four characters refer to another possible symmetry axis parallel to or tilted relative to the main translation axis: it can be characterized by a mirror plane (m), glide plane (g) or none (1) as before, or it can be characterized by the highest order of rotational symmetry n, which also determines the angle $\alpha = 180°/n$ that this last symmetry axis makes with respect to the main translation axis.

As examples, we consider two particular wallpaper groups. As shown in Fig. 4.3, the symmetry group $c2mm$ represents the common arrangement of bricks in structures with brick wall. The unit cell of the $c2mm$-group is shown in Fig. 4.4. The first character denotes a center unit cell, as is

Fig. 4.3   Example of the wallpaper group of *c2mm* in real life. Common household bricks possess the pattern symmetry of the wallpaper group *c2mm* in arrangement commonly found in brick walls. (Source: Wikipedia)

shown in Fig. 4.4. It has a rotational symmetry of order 2 such that the corresponding angle of rotation is 180°. There exists one such rotational element of order 2 whose center is on a reflection axis, and another whose center resides on an axis that does not contain the reflection element. The third and fourth characters in the Hermann-Mauguin notation indicates that there are mirror axes both perpendicular and parallel to the main translation axis. When there is no confusion, the full Hermann-Mauguin symbols may be shortened in appropriate and self-evident manner. As such, the *c2mm* group is often abbreviated as *cmm* in literature.

As another example, we consider the case of the wallpaper group denoted by *p1g1*. In contrast to the previous example of *c2mm* with its centered unit cell, the *p1g1*-wallpaper group is based upon repetition of

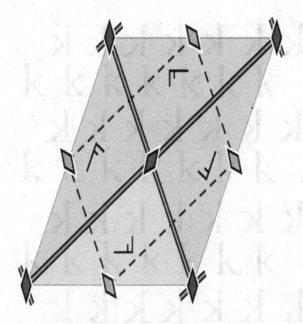

Fig. 4.4   Unit cell of *c2mm*. (Source: Wikipedia)

a primitive unit cell. A typical pattern with *p1g1*-symmetry is shown in Fig. 4.5, while its unit cell structure is shown in Fig. 4.6. This group contains glide reflections only, and their axes are all parallel. At times the wallpaper group of *p1g1* is abbreviated simply as the *pg*-group. In a similar fashion, the remaining 15 wallpaper groups follow similar logic for their notation. At the end of the chapter, we present a complete compilation of patterns that exemplifies the remaining 15 wallpaper symmetry groups (Figs. 4.7–4.22).

4.3.2 *Orbifold notation*

Another way to organize the various wallpaper groups in two dimensions is to recognize that two-dimensional tessellation may be represented as the mathematical operation of obtaining a quotient space of the two-dimensional Euclidean space with respect to a finite group (space group). The aforementioned quotient space is termed *orbifold*, whose study falls under the subject of topology and geometric group theory [29]. For completeness, we shall now define more rigorously the concepts of quotient spaces and manifolds as they pertain to the context of orbifolds.

k k k k k k k
ʞ ʞ ʞ ʞ ʞ ʞ ʞ
k k k k k k k
ʞ ʞ ʞ ʞ ʞ ʞ ʞ
k k k k k k k
ʞ ʞ ʞ ʞ ʞ ʞ ʞ
k k k k k k k
ʞ ʞ ʞ ʞ ʞ ʞ ʞ

Fig. 4.5 *p1g1*. In orbifold notation, *xx*. The above pattern is constructed from the repeated motif of the letter "k" shown in the lower right corner.

In physics, we mostly work with spaces that come equipped with a *metric*, i.e. a measure that gives the concept of distance in the given spaces. We will refer to such spaces endowed with a metric *metric space*. There are many types of mathematical spaces that one uses in physics to describe a number of different physical systems. Besides metric spaces, the concept of a *manifold* is often quite useful as a generalization to the familiar *Euclidean spaces* intuitive to our senses. A manifold is a more general mathematical structure such that for each point in this manifold there exists a

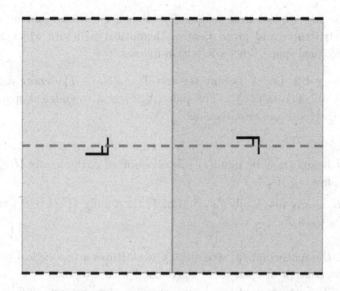

Fig. 4.6   Unit cell of $p1g1$. (Source: Wikipedia)

local neighborhood that resembles a Euclidean space but that globally can have the possibility of a more complicated structure. One such physical example would be the surface of the Earth which approximates the shape of a sphere. Manifolds are important since it allows us to apply well-known properties of Euclidean spaces to global structures more complicated than simple Euclidean spaces. However, a yet more general kind of mathematical structures exist that do not possess a metric at all, which nevertheless can be quite useful when describing physical systems in which the concept of distance is not particularly germane. Without the concept of distance measure, which is the basic assumption in *geometry* as its name implies, we arrive at the subject matter of *topology* that is a particular branch of geometry that deals with a class of transformations known as *homeomorphisms*. It builds upon set theory and considers both sets of points and families of sets. In this context, we generally refer to mathematical spaces without an endowed metric as *topological spaces*. To sum up, metric spaces form a subset of manifolds, while manifolds form a subset of topological spaces. As is conventionally done in mathematics, we start with the definition of the most general concept and precede to the more particular cases in the following.

We begin with the most general (one that has the least amount of pre-scribed structures and properties) mathematical structure which we will call topological space, whose definition follows:

Definition 4.2. Let X be any set and $\mathcal{T} = \{U_i | i \in I\}$ denote a certain collection of subsets of X. The pair $(X, \mathcal{T})$ is a topological space if $\mathcal{T}$ satisfies the following requirements:

 (i) $\{\emptyset, X\} \in \mathcal{T}$,

 (ii) If J is any (may be infinite) subcollection of I, the family $\{U_j | j \in J\}$ satisfies $\cup_{j \in J} U_j \in \mathcal{T}$.

 (iii) If K is any finite subcollection of I, the family $\{U_k | k \in K\}$ satisfies $\cap_{k \in K} U_k \in \mathcal{T}$.

As such, the mathematical structure X constitutes a topological space, in which the concept of open sets U_i is defined with a topology $\mathcal{T}$.

Thus, the structure of topological space is quite general, and it can be tailored to describe a variety of systems upon the endowment of additional structures to better adapt to a given physical system of interest. As previously mentioned, the concept of a distance measure is one such example of an additional useful structure to the topological space. The resulting space is known as a metric space. By definition,

Definition 4.3. A metric $d : X \times X \to \mathbb{R}$ is a function that satisfies the following:

 (i) $d(x, y) = d(y, x)$

 (ii) $d(x, y) \geq 0$ where the equality holds if and only if $x = y$.

 (iii) $d(x, y) + d(y, z) \geq d(x, z)$ for any $x, y, z \in X$.

When a topological space is endowed with a metric and thus becoming a metric space, we can further refine the concept of open sets into open discs by the following qualification $U_\epsilon(x) = \{y \in X | d(x, y) < \epsilon\}$.

Beyond the concept of distance, it is also convenient to endow the property of smoothness in a topological space so that we can develop the corresponding *calculus* for that space. A manifold is the consequent topological space endowed with the smoothness property, and we now present it definitely in a more precise way, as follows:

Fig. 4.7 *p*2. In orbifold notation, 2222. The above pattern is constructed from the repeated motif of the letter "k" shown in the lower right corner. We note that the motif itself has no nontrivial element of internal symmetry other than identity.

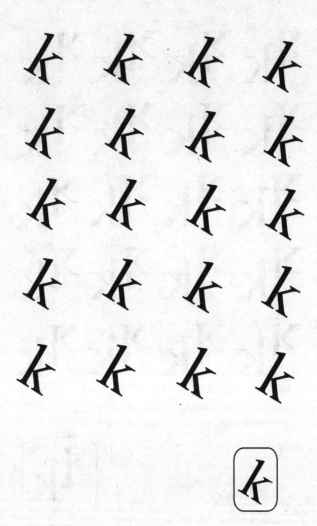

Fig. 4.8   *p*1. In orbifold notation, *o*. The above pattern is constructed from the repeated motif of the letter "k" shown in the lower right corner. We note that the motif itself has no nontrivial element of internal symmetry other than identity.

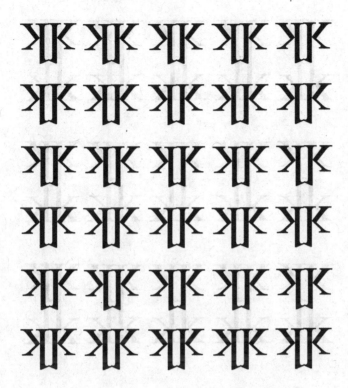

Fig. 4.9 *p1m1*. In orbifold notation, **. The above pattern is constructed from the repeated motif of the letter "k" shown in the lower right corner.

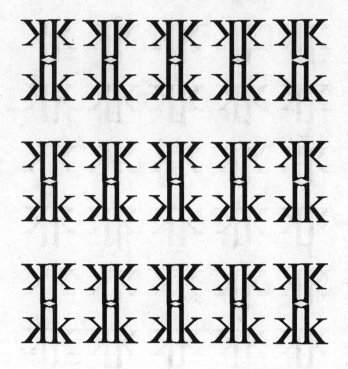

Fig. 4.10   *p2mm*. In orbifold notation, *2222. The above pattern is constructed from the repeated motif of the letter "k" shown in the lower right corner.

Fig. 4.11 *p2mg*. In orbifold notation, 22*. The above pattern is constructed from the repeated motif of the letter "k" shown in the lower right corner.

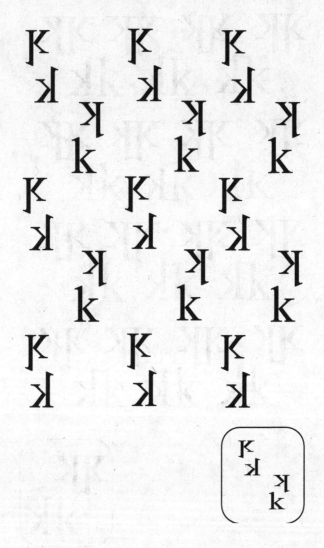

Fig. 4.12   *p2gg.* In orbifold notation, 22x. The above pattern is constructed from the repeated motif of the letter "k" shown in the lower right corner.

Fig. 4.13 *c1m1*. In orbifold notation, **x*. The above pattern is constructed from the repeated motif of the letter "k" shown in the lower right corner.

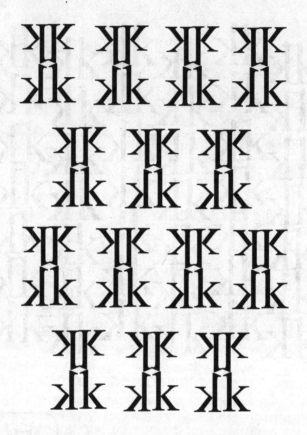

Fig. 4.14 *c2mm*. In orbifold notation, 2*22. The above pattern is constructed from the repeated motif of the letter "k".

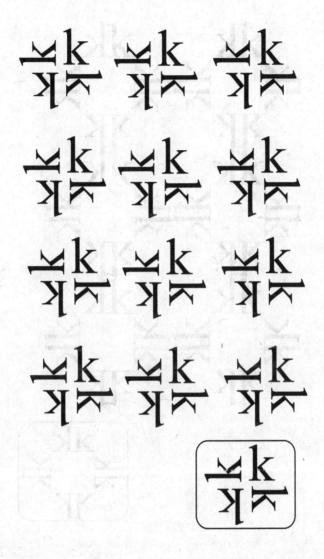

Fig. 4.15 *p*4. In orbifold notation, 442. The above pattern is constructed from the repeated motif of the letter "k".

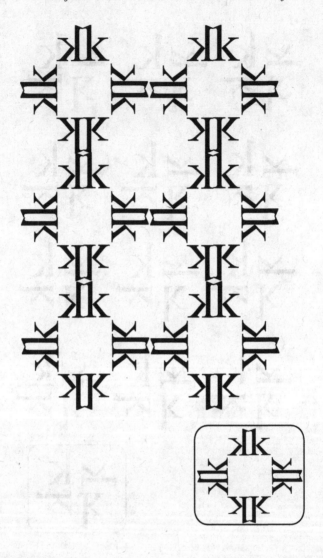

Fig. 4.16   *p4mm*. In orbifold notation, *442. The above pattern is constructed from the repeated motif of the letter "k".

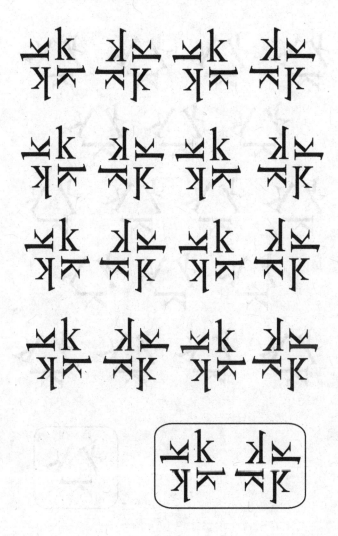

Fig. 4.17 *p4gm*. In orbifold notation, 4*2. The above pattern is constructed from the repeated motif of the letter "k".

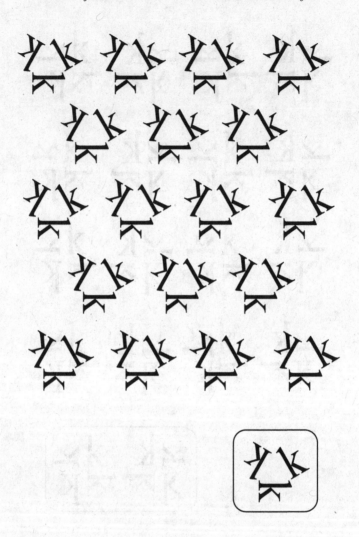

Fig. 4.18   *p*3. In orbifold notation, 333. The above pattern is constructed from the repeated motif of the letter "k".

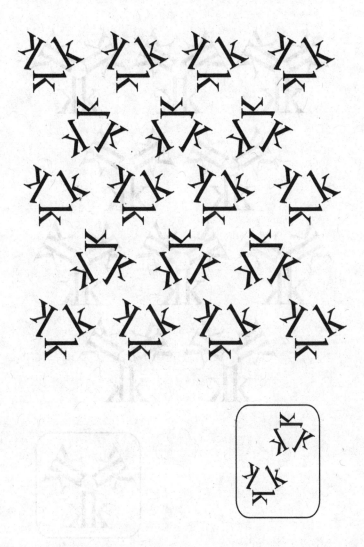

Fig. 4.19 *p3m1*. In orbifold notation, *333. The above pattern is constructed from the repeated motif of the letter "k".

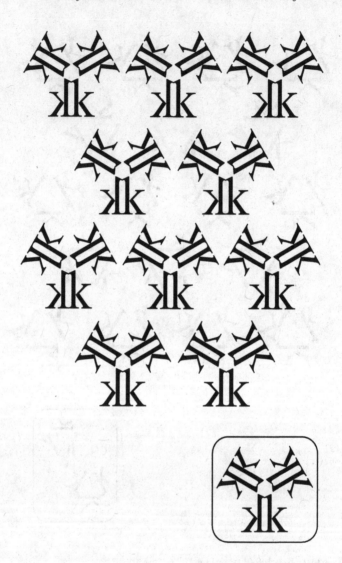

Fig. 4.20 $p31m$. In orbifold notation, 3*3. The above pattern is constructed from the repeated motif of the letter "k".

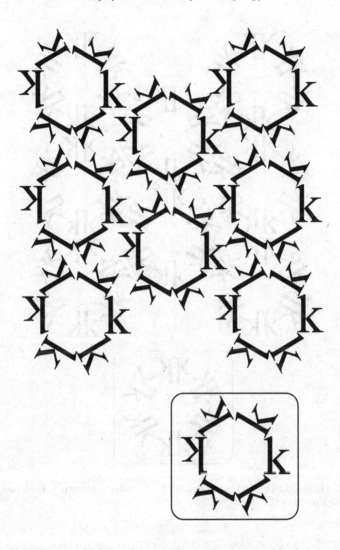

Fig. 4.21 *p*6. In orbifold notation, 632. The above pattern is constructed from the repeated motif of the letter "k".

Fig. 4.22 *p6mm*. In orbifold notation, *632. The above pattern is constructed from the repeated motif of the letter "k".

Definition 4.4. M is an m-dimensional differentiable manifold if

 (i) M is a topological space,

 (ii) M is provided with a family of pairs $\{(U_i, \varphi_i)\}$.

 (iii) $\{U_i\}$ is a family of open sets which covers M, that is, $\cap_i U_i = M$. φ_i is a homeomorphism from U_i onto an open subset U_i' of $\mathbb{R}^m$.

 (iv) Given U_i and U_j such that $U_i \cap U_j \neq \emptyset$, the map $\psi_{ij} = \varphi_i \varphi_j^{-1}$ from $\varphi_j (U_i \cup U_j)$ to $\varphi_i (U_i \cup U_j)$ is infinitely differentiable.

Thus, as much as topology is concerned about the property of continuity, manifolds are concerned about the smoothness property of a particular mathematical structure. Conceptually, a manifold can be thought of as a general mathematical space wherein we frame an appropriate coordinate system at each local neighborhood that enables us to utilize all the tools of linear algebra and calculus.

Going back to our system of two-dimensional lattices, we can think of them as examples of manifolds being divided up into subspaces that are identical and repetitive in nature. Mathematically, this action of "dividing" is furnished by an appropriate *mapping* between the total space and the individual subspaces. To begin, we consider the definition of a map:

Definition 4.5. A mapping is a rule by which we assign $y \in Y$ for each $x \in X$ and is denoted by $f : X \to Y$.

A related concept is that of *equivalence relations*, essentially identification maps, which can be defined as follows:

Definition 4.6. An equivalence relation $\sim$ is a relation which satisfies the following requirements:

(i) $a \sim a$.

(ii) If $a \sim b$, then $b \sim a$.

(iii) If $a \sim b$ and $b \sim c$, then $a \sim c$.

The three properties are generally referred to as *reflective, symemtric*, and *transitive* properties, respectively.

Given a set X and an equivalence relation $\sim$, we can have a partition of X into mutually disjoint subsets known as *equivalent classes*. In its simplest guise, a class $[w]$ consists of all the elements x in X such that $x \sim w$, as follows:

$$[w] \equiv \{x \in X | x \sim w\}. \tag{4.19}$$

The set of all equivalent classes is known as the *quotient space* and denoted by $X/\sim$ for a topological space X. Before proceeding further, we consider a simple illustration in which we apply the above language in describing certain simple geometrical objects possessing simple symmetries.

Let us for the moment consider a square disc X defined by $\{(x + 2\pi n_x, y + 2\pi n_y) \in \mathbb{R}^2 | n_x, n_t \in \mathbb{Z}\}$. If we propose the equivalence relation in identifying the points on a pair of facing edges, $(-1, y) \equiv (1, y)$, we

obtain the cylinder (Fig. 4.23). On the other hand, if we make the identification of $(-1, -y) \equiv (1, y)$, we then obtain a Möbius strip. If we consider a
slight generalization of the aforementioned square disc into one defined by
the periodic boundaries $\{(x + 2\pi n_x, y + 2\pi n_y) \in \mathcal{R}^2 | n_x, n_y \in \mathbb{Z}\}$, the iden-
tification of all points $(x, y) \equiv (x + 2\pi n_x, y + 2\pi y_n)$ where $n_x, n_y \in \mathbb{Z}$, we
obtain the topology of a torus T^2 instead (Fig. 4.23).

Thus, viewing a two-dimensional tessellation pattern as an orbifold provides another direct and elegant means of categorizing all possible structures allowed by symmetry constraints. In addition to the wallpaper groups,
the orbifold notation can be equally well applied to classifying other symmetry groups in simply-connected two-dimensional spaces of constant curvature. Examples include the analogue of the wallpaper groups in the Euclidean space to their counterparts on the sphere (S^2) and the hyperbolic
plane (H^2) [30].

Conceptually, the orbifold notation represents the logical listing of all
symmetry-generating elements of the group. As previously mentioned, the
isometries of the Euclidean plane falls into one of the four categories: translations, rotations, reflections, and glide reflections. Mainly, we denote by
an integer n as the generator of an n-fold rotation. In order to distinguish
between a cyclic element of pure rotation and dihedral elements consisting
of centers of rotation with mirrors through them, we order generators of the
cyclic kind before an asterisk $(*)$ while list the generators of the dihedral
elements after the asterisk. The symbol of a cross (x) indicates a glide reflection, while the symbol (o) denotes null symmetry for systems possessing
translation symmetry and nothing else.

For illustration purposes, we revisit the two aforementioned examples
of the *c2mm*- and *p1g1*-wallpaper groups. In terms of orbifold notation,
the *c2mm*-group would be denoted by 2*22, while *p1g1*-group is denoted
by xx, as there are exactly two glide reflections present in each unit cell of
the latter. For the illustration of each lattice types, we include both the
crystallographic notation and the orbifold notation for each lattice.

4.3.3   *Why Are There Exactly 17 Wallpaper Groups?*

One advantage for the orbifold notation lies in not only its ability in effective enumerating the exhaustive list of all 17 wallpaper groups but also the
elucidation of the underlying reason for this total number — in terms of the
topological invariant known as the *Euler characteristic* denoted by χ. It is

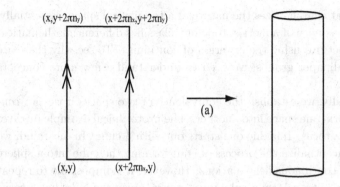

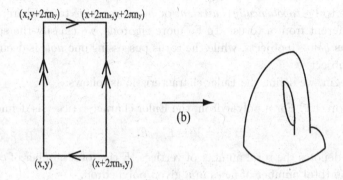

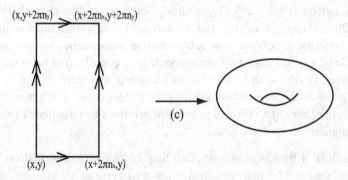

Fig. 4.23 Construction of cylinder (a), Möbius strip (b), and torus (c) based on different identifications (equivalence relations) on the edges of a square disc, parametrized by $\{(x + 2\pi n_x, y + 2\pi n_y) \in \mathbb{R}^2 | n_x, n_t \in \mathbb{Z}\}$.

a quantity that characterizes the nature of a topological space. Essentially, without the existence of a metric, topology classifies different mathematical spaces or structures using the criterion of continuity. To see why there are exactly 17 wallpaper groups, we need to understand a few more concepts from topology.

Conceptually, we establish the equivalence of two spaces if we can *continuously* deform one space into another. The prototypical example involves a sphere and a torus. Imagine one starts out with a cube. In our mind, we can easily conceptualize the process of transforming the cube into a sphere via a series of continuous deformations. However, it is impossible to repeat such a process to transform a cube into a torus, given the *hole* that a torus possesses. Therefore, in topology we would classify an object in the shape of a cube to be *topologically equivalent* or *homeomorphic* to a sphere while being different from a torus. To be more rigorous, we classify the sphere or cube as *genus*-0 objects, while the torus possessing one *hole* is deemed a *genus*-1 object.

To begin, we define the Euler characteristic as follows:

Definition 4.7. For a polyhedron, the Euler characterstics χ is defined as

$$\chi = V - E + F, \tag{4.20}$$

where V denotes the total number of vertices, E the total number of edges, and F the total number of faces in a given polyhedron.

For a simple demonstration, we consider the classical polyhedron of a dodecahedron (Fig. 4.24). By simple counting, we see that there are in total 20 vertices, 30 edges, and 12 faces. Using the definition of Euler characteristics, we obtain $\chi = 2$. As another illustration, we return to the example of a cube. Again by simple counting, we tally up a total number of 8 vertices, 12 edges, and 6 faces to again obtain the value of $\chi = 2$. Therefore, the dodecahedron is homeomorphic to the cube. For generalizations to objects other than classical polyhedra, we need to introduce the concept of *triangulations* and *simplexes*:

Definition 4.8. *Simplexes* are building blocks of a polyhedron. A 0-simplex, denoted by $\langle p_0 \rangle$, is simply a point or vertex. A 1-simplex, denoted by $\langle p_0 p_1 \rangle$ is a line or an edge joining the points p_0 and p_1. Analogously, a 2-simplex $\langle p_0 p_1 p_2 \rangle$ is defined to be the triangle bounded by the three points including its interior. Lastly, a 3-simplex is defined as a solid tetrahedron denoted by $\langle p_0 p_1 p_2 p_3 \rangle$.

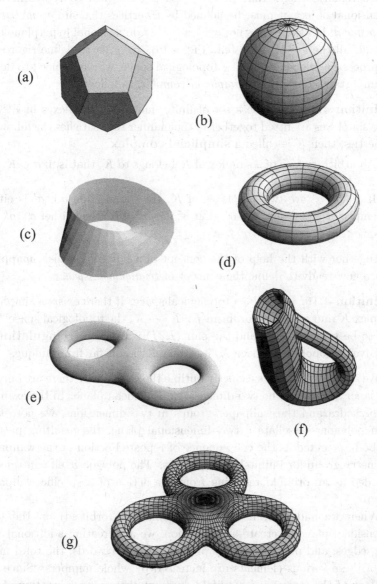

Fig. 4.24 Euler characteristics for different surfaces. Dodecahedron, $\chi = 2$ (a). Sphere, $\chi = 2$ (b). Möbius strip, $\chi = 0$ (c). Torus, $\chi = 0$ (d). Double torus, $\chi = -2$ (e). Klein's bottle, $\chi = 0$ (f). Triple torus, $\chi = -4$ (g). Figures are obtained using Mathematica (http://www.wolfram.com/mathematica).

We remark briefly that in order for a r-complex to represent a r-dimensional object, it must be defined by r vertices that are *geometrically independent*; i.e. there does not exist a $(r-1)$-dimensional hyperplane that contains all the $r+1$ points (c.f. Fig. 4.25). In order to define rigorously the process of triangulation of a topological space, we would need to define the important concept of a *simplicial complex*, as follows.

Definition 4.9. Let K be a set of finite number of simplexes in $\mathbb{R}^m$. If these simplexes are fitted together in the manner that satisfies the following properties, then K is called a **simplicial complex**.

(i) An arbitrary face of a simplex of K belongs to K, that is, if $\sigma \in K$ and $\sigma' \leq \sigma$, then $\sigma' \in K$.

(ii) If σ and σ' are two simplexes of K, the intersection $\sigma \cup \sigma'$ is either empty or a *face* of σ and σ', that is, if $\sigma, \sigma' \in K$, then either $\sigma \cap \sigma' = \emptyset$ or $\sigma \cap \sigma' \leq \sigma$ and $\sigma \cap \sigma' \leq \sigma'$.

Together with the help of the concept of a homeomorphism mapping, we are now ready to define the concept of *triangulable spaces*:

Definition 4.10. Let X be a topological space. If there exists a simplicial complex K and a homeomorphism $f : |K| \to X$, the topological space X is said to be **triangulable** and the pair (K, f) is called a **triangulation** of X. Given a topological space X, its triangulation is far from unique.

We are now ready to succinctly outline the reason why there are only 17 wallpaper groups for the two-dimensional Euclidean plane. In the example of polyhedra and the wallpaper groups in two dimensions, we note that when polygons tessellate a two-dimensional plane, the resulting pattern can be interpreted as the consequence of repeated action of the wallpaper symmetry group for infinitely many times. The polygon itself can thus be regarded as an orbifold resulting from the action of a specific wallpaper group.

When we add each new repeated unit cell (orbifold) on the two-dimensional plane to create a tessellation, we are creating additional vertices, edges, and polygons in the process. In other words, the total number of these various elements are increased by whole numbers. Since the topology of the underlying manifold does not change, each creation of the unit cell with all its vertices, edges, and polygons must maintain the Euler characteristic χ (Definition 4.7) before and after the process. We can also characterize the reverse of the above process of tessellation by assigning

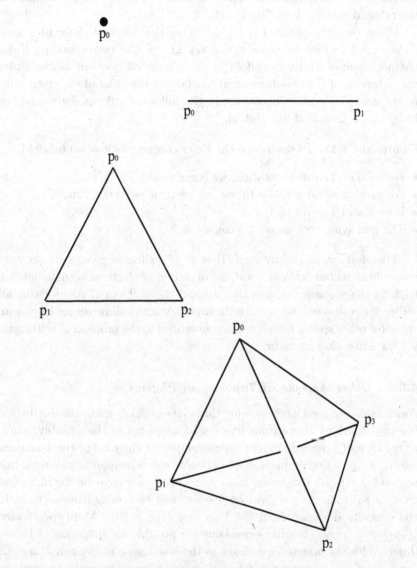

Fig. 4.25 Examples of simplexes. A simple geometric point corresponds to a 0-simplex (p_0). A line corresponds to a 1-simplex $\langle p_0 p_1 \rangle$. The mathematical object of a 2-simplex is denoted by $\langle p_0 p_1 p_2 \rangle$. The tetrahedron is an example of a 3-simplex denoted by $\langle p_0 p_1 p_2 p_3 \rangle$.

fractional numbers, instead of whole numbers, to these various elements of the orbifold under deconstruction [31].

Given that the orbifold is by definition the quotient space of a two-dimensional surface by some symmetry group, the corresponding Euler characteristic of a given orbifold is the associated quotient of the Euler characteristic of the two-dimensional surface by the order of the given symmetry group. Specifically, we apply the following rule to determine the Euler characteristic of an orbifold:

Definition 4.11. To determine the Euler characteristic of an orbifold:

- For each cyclic n-fold rotation, we count it as $(n-1)/n$.
- For each dihedral n-fold rotation, we count it as $(n-1)/2n$.
- Both $*$ and x count as 1.
- The null symmetry element o counts as 2.

Therefore, we can easily verify that all 17 wallpaper groups are the only possibilities whose orbifold notations of strings of elements actually add up to 2. In the process, we have also reduced the problem of enumerating all wallpaper groups consistent with the topology of two-dimensional Euclidean space (or other spaces with constant curvature) to the problem of arithmetic via the Euler characteristic.

4.3.4 *Other Aspects of Topology in Physics*

As an aside, we mention in passing that one can easily make the distinction in our definitions above from *unoriented simplexes* to the suitably generalized case of *oriented simplex*. For example, in the case of the 1-simplex $\langle p_0 p_1 \rangle$, a.k.a. a line segment, we can denote the corresponding *directed* line segment as $(p_0 p_1)$ traversing from p_0 to p_1, which is to be distinguished from $(p_1 p_0)$ indicating a directed line segment traversing from p_1 to p_0 in the opposite direction from the first case (Fig. 4.25). More specifically, $(p_0 p_1) = -(p_1 p_0)$. Similar extensions are possible for simplexes of higher order. With the oriented simplexes as the starting point, we can define the corresponding *cycle group* and *boundary group*, which form the basis of the important tool of *homology group* in classifying topological spaces based on the algebraic structures they are endowed with [32]. Similar algebraic structures can be introduced in the study of continuous deformations of maps, which underlies the study of *homotopy groups* whose applications include its use in classifying the structure of defects in many condensed-matter systems such as liquid-crystals, lattices, and superconducting vortices [33]. Lastly, the concept of *duality* and the study of the *de Rham cohomology group* as the dual of the aforementioned homology group has also proved useful in many areas of theoretical physics that make use of *differential forms*.

4.4 Three-Dimensional Point Groups

In three dimensions, there are vastly more space groups that is consistent with three-dimensional translational symmetry. There are in total 230 space groups in three-dimensional Euclidean space. Much like the five basic lattice types classified previously for the two-dimensional case, we can similarly start with the classification of 14 possible Bravais lattices: triclinic, monoclinic, orthorhombic, trigonal, hexagonal, tetragonal, and isomeric in varieties of simple, face-centered, and body-centered lattices (Fig. 4.26). Combined with the 32 possible point groups, along with the possibility of *screw axis* and the different orientation of the glide and translation directions, we can enumerate in total 230 space groups in three dimensions. For a complete compilation of all space groups, we refer the reader to the many excellent existing references [34].

In what follows, we will be particularly interested in the cubic lattices of the face-centered (FCC) and body-centered (BCC) varieties as well as a non-closed packed structure known as A15 that possesses the symmetry of $Pm\bar{3}n$ in crystallographic notation. While the reasons behind the formation of the first two lattice types, FCC and BCC lattices, are well-known and accepted in the condensed-matter physics community, the reason for the exotic $A15$-lattice is not yet clear. It is the purpose of our model to explain its realization in actual physical systems of colloids. More specifically, we explain its origin in terms of packing geometry and the associated energetics in crystal lattices.

4.4.1 *Face-centered Cubic (FCC) Lattices*

The FCC lattice takes on the crystallographic symbol of $Fm\bar{3}m$. Common metallic elements that take on this structure include aluminum (Al), copper (Cu), nickel (Ni), strontium (Sr), silver (Ag), gold (Au), and lead (Pb). For the FCC lattice, each atom in the corner is shared by eight other unit cells, while a face-centered atom is shared by only two. Therefore, there are a

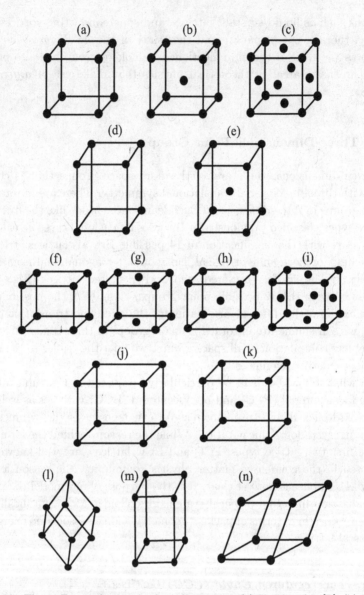

Fig. 4.26   The 14 Bravais lattices in three dimensions. (a) Simple cubic [P]. (b) Body-centered cubic [I]. (c) Face-centered cubic [F]. (d) Simple tetragonal [P]. (e) Body-centered tetragonal [I]. (f) Orthorhombic [P]. (g) Orthorhombic [C]. (h) Orthorhombic [I]. (i) Orthorhombic [F]. (j) Monoclinic [P]. (k) Monoclinic [C]. (l) Trigonal [R]. (m) Hexagonal [P]. (n) Triclinic [P].

total of four atoms per unit cell for the FCC lattice. The Wigner-Seitz cell of the FCC lattice is the rhombic dodecahedron. The unit cell of the FCC lattice is depicted in Fig. 4.27.

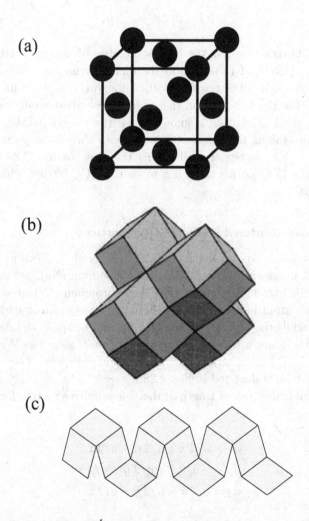

(a)

(b)

(c)

Fig. 4.27 Unit cell of the FCC lattice (a) and the corresponding three-dimensional Wigner-Seitz cells with the geometry of rhombic dodecahedron (b). The net image of a unit cell of the rhombic dodecahedron (c). (Source: Kamien group, University of Pennsylvania, Philadelphia.)

For a cubic unit cell of length a, the non-primitive vectors for the FCC lattice are

$$a_1 = a/2\,\hat{y} + a/2\,\hat{z}\,,$$
$$a_2 = a/2\,\hat{x} + a/2\,\hat{z}\,,$$
$$a_3 = a/2\,\hat{x} + a/2\,\hat{y}\,. \tag{4.21}$$

The FCC structure has the highest number of nearest neighbors (co-ordination number) of 12 as well as the highest symmetry. While its conventional unit cell takes the form of a cube, we note that its primitive cell is not. For the same reason that it has the highest number of nearest neighbors, the FCC lattice is known to be the closest packing structure for molecules taking the shape of spheres, with the closest-packed volume fraction $\phi \approx 0.74$ for spheres all having the same radius. The reciprocal lattice of the FCC lattice turns out to be the BCC lattice, which we will discuss next.

4.4.2   *Body-Centered Cubic (BCC) Lattices*

The BCC lattice takes on the crystallographic symbol of $Im\bar{3}m$. Common elements that take on this structure include sodium (Na), potassium (K), cesium (CS), barium (Ba), iron (Fe), and chromium (Cr). For the BCC lattice, each atom in the corner of the unit cell is again shared by other unit cells, while the body-centered atom belongs uniquely to one.

Thus, there are a total of two atoms per unit cell. The Wigner-Seitz cell of the BCC lattice is the Kelvin's tetrakaidecahedron. The unit cell of the BCC lattice is depicted in Fig. 4.28.

For a cubic unit cell of length a, the non-primitive vectors for the BCC lattice are

$$a_1 = a/2\,\hat{x} + a/2\,\hat{y} + a/2\,\hat{z}\,,$$
$$a_2 = -a/2\,\hat{x} + a/2\,\hat{y} + a/2\,\hat{z}\,,$$
$$a_3 = a/2\,\hat{x} - a/2\,\hat{y} + a/2\,\hat{z}\,. \tag{4.22}$$

The BCC structure has eight nearest neighbors for each particle, and the packing efficiency of this lattice is about 0.68 when compared with 0.74 for the FCC lattice. Incidentally, the reciprocal lattice of the BCC lattice is the FCC lattice.

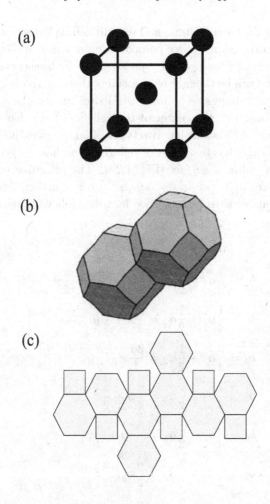

Fig. 4.28 Unit cell of the BCC lattice (a) and the corresponding three-dimensional Wigner-Seitz cells with the geometry of Kelvin's tetrakaidecahedron (b). The net image of a unit cell of the Kelvin's tetrakaidecadedron (c). (Source: Kamien group, University of Pennsylvania, Philadelphia.)

4.4.3 *A15 Lattices*

The A15 lattice takes on the crystallographic symbol of $pm\bar{3}n$. Common compounds that take on this structure include β-tungsten, Nb_3Al, and Ti_3Sb. Unlike the aforementioned two cases of the FCC and BCC lattices, the A15 lattice consists of two different types of lattice sites. This lattice also belongs to the cubic system, and its unit cell includes eight sites which can be further divided into three pairs of columnar sites and two interstitial sites. The columnar sites lie evenly spaced along the bisectors of the faces of the unit cell and can be thought of as forming three mutually perpendicular and interlocking columns. The interstitial sites fill out the space between the columns: one is at the center of the cell, and the other one is at the vertex. Since the A15 lattice has two types of lattice sites, its Wigner-Seitz cell correspondingly has two types of cell units and has the geometry of the *Weaire-Phelan minimal surface* (Fig. 4.29). The primitive vectors for the A15 unit cell are $a_1 = a\,\hat{x}$, $a_2 = a\,\hat{y}$, $a_3 = a\,\hat{z}$. Furthermore, due to the two kinds of lattice sites possible, we have the following basis vectors for the A15 lattice:

$$b_1 = 0$$

$$b_2 = \frac{1}{2}a_1 + \frac{1}{2}a_2 + \frac{1}{2}a_3 = \frac{a}{2}\hat{x} + \frac{a}{2}\hat{y} + \frac{a}{2}\hat{z}$$

$$b_3 = \frac{1}{4}a_1 + \frac{1}{2}a_2 = \frac{a}{4}\hat{x} + \frac{a}{2}\hat{y}$$

$$b_4 = \frac{3}{4}a_1 + \frac{1}{2}a_2 = \frac{3a}{4}\hat{x} + \frac{a}{2}\hat{y}$$

$$b_5 = \frac{1}{4}a_2 + \frac{1}{2}a_3 = \frac{a}{4}\hat{y} + \frac{a}{2}\hat{z}$$

$$b_6 = \frac{3}{4}a_2 + \frac{1}{2}a_3 = \frac{3a}{4}\hat{y} + \frac{a}{2}\hat{z}$$

$$b_7 = \frac{1}{2}a_1 + \frac{1}{2}a_3 = \frac{a}{2}\hat{x} + \frac{a}{2}\hat{z}$$

$$b_7 = \frac{1}{2}a_1 + \frac{3}{4}a_3 = \frac{a}{2}\hat{x} + \frac{3a}{4}\hat{z} \qquad (4.23)$$

The A15 lattice as well as its Wigner-Seitz cell are depicted in Fig. 4.29. The special mathematical property of its Wigner-Seitz cell would explain its observation in crystals of self-assembled micelles and of some dendritic polymers [35], as we will detail in the following sections.

4.5   Conceptual Framework of the Foam Model

Now we make use of the fact in our specific context that the particles are themselves colloids, which can be modeled mathematically as hard spheres.

As previously mentioned, we will consider two specific classes of colloidal particles: charged and fuzzy colloids. In either case when we consider the whole system at fixed density, the total volume of the system is the sum of the volume of the hard cores and the excess volume of either the salt solution in charged systems or the soft chains within the corona in fuzzy colloids (c.f. Fig. 4.30). Since this latter volume can be viewed as enveloping the individual spheres, we imagine breaking the volume up into a lattice of Voronoi cells, each of which contains a colloidal particle. The excess volume can then be written, by definition, as the product of the area of these dividing surfaces and their average thickness. Since the volume of the hard cores and the volume of the whole system is fixed at a given density, the excess volume is also constant and so:

$$A_m d = \text{constant}, \tag{4.24}$$

where A_m is the total area of these bilayers and d is their average thickness. The excess volume depends on the particle density of the system. Using simple geometrical considerations, the volume per particle subtracting the volume of each particle is:

$$A_m d = 2 \left(\frac{1}{n} - \frac{\pi}{6} \right) \sigma^3, \tag{4.25}$$

where $A_m = \gamma^x \sigma^2 n^{-2/3}$, depends on the lattice type ($\gamma^{FCC} = 5.345$, $\gamma^{BCC} = 5.308, \gamma^{A15} = 5.288$) and is a dimensionless quantity characterizing the magnitude of the ratio of surface area to volume per cell.

With our geometrical expressions in hand for the three lattices, we can now freely utilize them in the construction of the free energy density for each of these lattices. As expected, they would encode geometric information particular to the distinct geometry of each lattice candidate. We will continue in the next chapter to discuss the physical ramifications of such free-energy construction, based on the foam model, and explore the consequent thermodynamics properties and phase stabilities in subsequent chapters. To conclude our current survey of the important mathematical topics behind the foam model, we will now discuss a mathematical problem, first posed by Lord Kelvin and subsequently explored by a few other mathematicians that included Johannes Kepler, that served as an inspiration of our foam model.

(a)

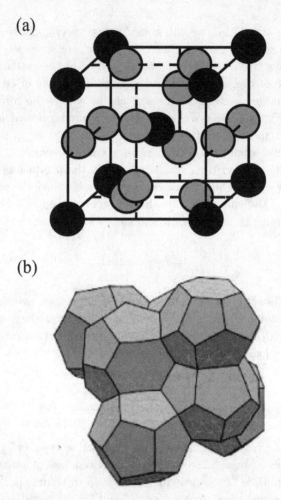

(b)

Fig. 4.29 Unit cell of the A15 lattice (a) and the corresponding three-dimensional Wigner-Seitz cells with the geometry of Weaire-Phelan minimal surface (b). (Source: Kamien group, University of Pennsylvania, Philadelphia.)

4.6 The Kelvin Problem and the Kepler Conjecture

The construction of our foam analogy depends on the partitioning of a finite volume into units of repeating cells. These cell units, in additional, are to be space-filling, i.e. they can generate a tessellation of space. In some instances, the interaction between the colloidal particles physically engaged within these repeating cells would require the optimal cell structure to

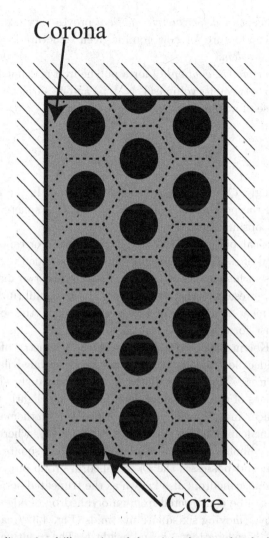

Fig. 4.30 A two-dimensional illustration of the minimal-area rule: the hard cores are embedded in the matrix of coronas, and the volume of the latter is given by the product of its area (dotted line) and the average separation of the cores.

have minimal surface area. The mathematical problem of determining a space-filling structure based on similar cell types of equal volume that also minimizes the corresponding surface area of the overall structure was first posed by Lord Kelvin in 1887; the original mathematical problem arose in the context of light propagation in a crystal [2]. In the following discussion, the concept of *isoperimetric quotient* proves very useful in providing a quantitative measure for how good a given structure is as a solution to the *Kelvin's problem*.

The *isoperimetric quotient* Q for a polyhedron is defined as the dimensionless ratio between its volume ($V = \frac{4}{3}r_V^3$) and its surface area ($S = 4\pi r_A^2$) using the case of a sphere as the reference point:

$$Q \equiv \frac{r_V^2}{r_A^3} = \frac{36\pi V^2}{A^3}. \qquad (4.26)$$

Not surprisingly, Q must take values less than one. In the limit of $Q \to 1$, the shape of the polyhedron converges to that of a sphere, which has the theoretically optimal ratio of surface area to volume.

The physical picture of the Kelvin's problem corresponds exactly to the formation of soap films. Joseph Plateau expounded upon the stability of soap films in his classic 1873 book: he proposed two main rules whose satisfaction represent mechanical stability. In an equilibrium froth, the adjacent faces meet at an angle of $120°$ and the edges (the so-called Plateau borders) must form a tetrahedral angle of $109°28'$. Based upon experimentation, Kelvin proposed a solution in the geometry of a 14-sided orthic tetrakaidecahedron (Fig. 4.31) consisting of six quadrilateral and eight hexagonal faces. More particularly, in order to satisfy the two *Plateau rules*, the edges of the tetrakaidecahedron must be slightly curved and the hexagonal faces must be somewhat non-planar. Mathematically, the 14-faced Archimedean solid A_{12} has faces $8\{6\} + 6\{4\}$, where we have used the Schläfli symbols denoting a regular hexagon and square respectively by $\{6\}$ and $\{4\}$, and possesses the O_h octahedral group of symmetries. The orthic tetrakaidecahedron, also known as the truncated octahedron, of unit edge length can be constructed from an octahedron of edge length of 3 via truncation by removing six square pyramids (Fig. 4.32), each with a slant height $s = 1$, base length $a = 1$, and height h. The resulting surface area A and volume V of the orthic tetrakaidecahedron can then be calculated in a straightforward manner:

$$S = 6 + 12\sqrt{3}, \qquad (4.27)$$

$$V = V_{\text{octahedron}} - 6\left(\frac{1}{3}A_b\,h\right), \qquad (4.28)$$

$$= 9\sqrt{2} - \sqrt{2} = 8\sqrt{2}.$$

The isoperimetric quotient Q of the orthic tetrakaidecahedron is:

$$Q = \frac{64\pi}{3(1 + 2\sqrt{3})^3},$$ (4.29)

$$\approx 0.753367,$$ (4.30)

versus $Q \approx 0.757$ for Kelvin's slightly curved variant of the polyhedron. In our context, these polyhedra turn out to be the Voronoi cell of the BCC lattice. Consequently, while the BCC lattice does not have the greatest packing density as the FCC lattice does, its Voronoi cells have smaller surface area in general than that of the FCC lattice — the rhombic dodecahedron.

While Kelvin's problem seems easy to state and hard to solve, a rigorous proof to the fact that FCC packing has the highest density was equally daunting. To the mathematics community, the solution was long known as the *Kepler conjecture*. Kepler stated that no packing of balls of the same radii in three dimensions has density greater than the face-centered cubic packing. The Kepler conjecture, in fact, was the 18th of the 23 Hilbert problems posed at the end of the 20th century by famed mathematician David Hilbert. The conjecture was confirmed nearly by Thomas Hales in 1998 [13], some 400 years later after Kepler had first made his conjecture.

Unlike the case of the Kepler conjecture which was rigorously confirmed, the Kelvin conjecture was disproved on the other hand. In 1994, Denis Weaire and Robert Phelan produced a foam structure with the symmetry of the A15 lattice (c.f. 3.4.3) with cavities of equal volume and surface area much smaller than the Kelvin structure of orthic tetrakaidecahedron. More specifically, the A15 lattice has two different types of lattice sites, and the A15 foam consists of 6 Goldberg tetrakaidecahedra, each with 2 hexagonal and 12 pentagonal faces that form 3 sets of interlocking columns, as well as two irregular pentagonal dodecahedra at the interstices (Fig. 4.29). The A15 lattice is derived from the polytope $\{3, 5, 5\}$, and only the hexagonal faces in this structure remain planar.

To elaborate further, a *polytope* is a generalization of the concept of polygons. In particular, while we define a *regular polygon* as a polygon whose edges and angles are both respectively all equal, and we define a regular polyhedron as having all faces being congruent regular polygons with all congruent and regular vertex figures, we similarly define a *regular n-polytope* as an n-dimensional polytope whose $(n - 1)$-dimensional faces are all regular and congruent, with the same rule of regularity and congruence applied to all vertex figures contained therein. A consistent way of

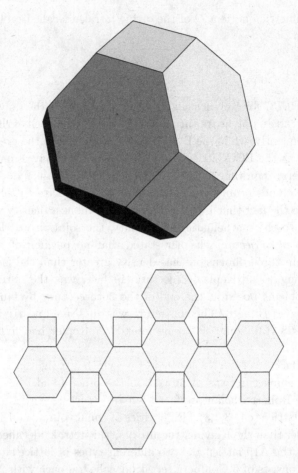

Fig. 4.31 Truncated Octehedron. The geometric structure that solves Kelvin's conjecture in finding the ideal configuration of a soap froth. Figures are obtained using Mathematica (http://www.wolfram.com/mathematica).

classifying and denoting polytopes is through the use of Schläfli symbols, which we have previously seen for its usage in denoting regular polygons. The generalization to higher dimensions is straightforward. For example, for a regular polyhedron having faces $\{n\}$ with p faces joining around a vertex, the Schläfli symbol would be $\{n, p\}$. The nine regular polyhedra in three dimensions can be numerated as $\{3, 3\}$, $\{3, 4\}$, $\{4, 3\}$, $\{3, 5\}$, $\{5, 3\}$, $\{5, 5/2\}$, $\{5/2, 5\}$, $\{3, 5/2\}$, and $\{5/2, 3\}$. Likewise, a regular *polycell* having cells $\{n, p\}$ with q cells joining around a vertex is denoted by $\{n, p, q\}$,

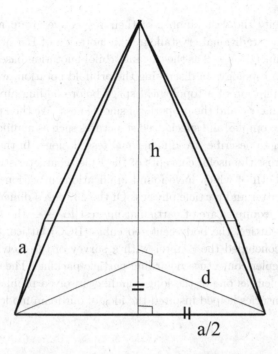

Fig. 4.32   Measurements of the square pyramid truncated from the octahedron in orthic tetrakaidecahedron; $d = \frac{\sqrt{2}}{2}a$.

of which the vertex figure is a $\{p, q\}$. The notation of polytope type of $\{3, 5, 5\}$ for the A15 lattice is now self-explanatory.

Incidentally, the A15 foam has an isoperimetric quotient $Q = 0.764$, which is approximately 0.3% less than that of the orthic tetrakaidecahedron in the Kelvin conjecture. It remains to be seen whether there will be other structures possessing an even smaller surface area per unit volume than the A15 foam; the proposed Weaire-Phelan solution remains only a conjecture at this point.

4.7   Conclusion

In this chapter, we have provided a survey of the pertinent mathematical concepts behind the premise of our foam model. We started out with the mathematical definition of lattices based on symmetry groups. We then proceeded to delineate the relevant aspects of the two-dimensional wall-paper groups and the three-dimensional space groups. To succinctly and logically classify the vast number of their respective group members, we introduced the traditional crystallographic notation of Hermann-Mauguin based on symmetry as well as the elegant orbifold notation based on modern concepts from topology. In discussing the orbifold notation, we introduced the bare assumption of a topological space before adding auxiliary structures such as metric and the property of smoothness. We then progressively built up more complex and specialized structures such as manifolds and orbifolds in order to describe two-dimensional tessellations. In the process, we also came across the useful concepts of the Euler characteristic and homotopy group, both of which have found applications in various branches of condensed-matter and particle physics. Of the 237 three-dimensional space groups, three examples are of particular interest to us — the face-centered cubic (FCC) lattice, the body-centered cubic (BCC) lattice, and the A15 lattice. We concluded the chapter with a survey of the Kelvin's problem and of the Kepler conjecture regarding particle packing. The status of the Kelvin's problem as one of the long-standing unsolved problem in mathematics for centuries helped inspired the idea of our foam model in colloidal physics.

References

[1] P. Ziherl and R. D. Kamien, *Phys. Rev. Lett.* **85**, 3528 (2000).

[2] P. Ziherl and R. D. Kamien, *J. Phys. Chem. B***105**, 10147 (2001).

[3] W. Kung, P. Ziherl and R. D. Kamien, *Phys. Rev. E* **65**, 050401(R) (2002).

[4] R. L. Whetten, M. N. Shafigullin, J. T. Khoury, T. G. Schaaff, I. Vezmar, M. M. Alvarez and A. Wilkinson, *Acc. Chem. Res.* **32**, 397 (1999).

[5] E. B. Sirota, H. D. Ou-Yang, S. K. Sinha, P. M. Chaikin, J. D. Axe and Y. Fujii, *Phys. Rev. Lett.* **62**, 1524 (1989).

[6] W. A. Curtin and K. Runge, *Phys. Rev. A* **35**, 4755 (1987).

[7] A. R. Denton and N. W. Ashcroft, *Phys. Rev. A* **39**, 4701 (1989).

[8] G. A. McConnell and A. P. Gast, *Phys. Rev. E* **54**, 5448 (1996).

[9] T. L. Hill, *Statistical Mechanics*, McGraw-Hill, New York, 1956.

[10] J. A. Barker, *Lattice Theory of the Liquid State*, Pergamon Press, Oxford, 1963.

[11] We thank M. O. Robbins for discussions on this point.

[12] V. S. K. Balagurusamy, G. Ungar, V. Percec and G. Johansson, *J. Am. Chem. Soc.*, **119**, 1539 (1997).

[13] T. C. Hales, *Notices Amer. Math. Soc.*, **47**, 440 (2000).

[14] W. Kohn and L. J. Sham, *Physical Review*, **140**, 1133A 1965.

[15] J. Z. Wu, *AlChE J.* **52**, 1169 (2006).

[16] M. Rex, H. H. Wensink, and H. Lowen, *Phys. Rev. E* **76**, 021403 (2007).

[17] J. Kleis, B. L. Lundqvist, D. C. Langreth, and E. Schroder, *Phys. Rev. B* **76**, 100201(R) (2007).

[18] P. C. Hiemenz and R. Rajagopalan, *Principles of Colloid and Surface Chemistry*, CRC, New York, 1997.

[19] B. Mennucci and R. Cammi, *Continuum Solvation Models in Chemical Physics: From Theory to Applications*, Wiley Press, New York, 2008; F. Hirata, *Molecular Theory of Solvation*, Springer Verlag, New York, 2003.

[20] P. G. de Gennes and J. Prost, *The Physics of Liquid Crystals*, Oxford University Press, Oxford, 1995; P. J. Collings, *Liquid Crystals: Nature's Delicate Phase of Matter*, Princeton University Press, Princeton, 2001.

[21] W. B. Russel, D. A. Saville, and W. R. Schowalter, *Colloidal Dispersions*, Cambridge University Press, Cambridge, 1992; R. J. Hunter, *Foundations of Colloid Science*, Oxford University Press, Oxford, 2001.

[22] T. V. Ramakrishnan and M. Yussouff, *Phys. Rev. B* **19**, 2775 (1979).

[23] W. A. Curtin and N. W. Ashcroft, *Phys. Rev. A* **32**, 2909 (1985).

[24] P. M. Chaikin and T. C. Lubensky, *Principles of Condensed Matter Physics*, Cambridge University Press, New York, 2000.

[25] W. D. Callister Jr., *Materials Science and Engineering: An Introduction*, John Wiley & Sons, New Jersey, 2003.

[26] J. Bamberg, G. Cairns and D. Kilminster, *American Mathematical Monthly* **110** (3), 202–209 (2003).

[27] S. A. Robertson, *Polytopes and Symmetry*, Cambridge University Press, Cambridge, 1984.

[28] H. Wondratschek and U. Muller, *International Tables for Crystallography, Volume A1: Symmetry Relations between Space Groups*, Springer Verlag, New York, 2004.

[29] B. A. Dubrovin, A. T. Fomenko, S. P. Novikov, and R. G. Burns, *Modern Goemetry: Methods and Applications: Part 2: The Geometry and Topology of Manifolds*, Springer Verlag, New York, 1985.

[30] E. B. Vinberg, *Geometry II: Spaces of Constant Curvature*, Springer Verlag, New York, 1993.

[31] A. Adem, J. Leida, and Y. Ruan, *Orbifolds and Stringy Topology*, Cambridge University Press, Cambridge, 2007; G. E. Martin, *Transformation Geometry: An Introduction to Symmetry*, Springer Verlag, New York, 1996.

[32] For details on these more advanced topics, readers are encouraged to consult M. Nakahara, *Geometry, Topology, and Physics*, Institute of Physics Publishing, Philadelphia, 1996.

[33] M. S. Dresselhaus, G. Dresselhaus, and A. Jorio, *Group Theory: Application to the Physics of Condensed Matter*, Springer Verlag, New York, 2008.

[34] T. Hahn, *International Tables for Crystallography, Volume A: Space Group Symmetry*, Springer Verlag, New York, 2002.

[35] A. Authier, *International Tables for Crystallography, Volume D: Physical Properties of Crystals*, Springer Verlag, New York, 2003.

PART 3
Geometry and Phase Transitions, in Colloidal Crystals

Chapter 5

Lattice Free Energy via the Foam Model

5.1 Introduction

In this work, we mainly consider systems of fuzzy colloids and charged colloids. Given the repulsive nature of both types of systems, the surfaces of the hard cores within each Voronoi cell would like to be as far apart from one another as possible. In light of our discussion on the conceptual framework of the foam model in Section 4.5, we see that maximizing the average thickness d in order to reduce the repulsive interaction would amount to minimizing the total interfacial area of the bilayers A_m. In other words, it follows that the *soft repulsion favors lattices whose Voronoi cells have smaller individual surface area.*

In addition to the soft repulsion that favors geometry of minimal surface area, there is the usual entropic considerations based on the orientation of individual colloidal particles subjected to the background thermal excitation. In principle, *the thermodynamically favored configuration should maximize the overall entropy (packing density) of the system.* It is this inherent frustration between entropy and repulsive interaction that underlies the variety of lattices, close-packed as well as non-close packed structures, that we observe in experiments for the solid phase of colloidal crystals.

By construction in our foam model, the total free energy can be divided into two contributions: a bulk part due to the positional entropy of the hard cores in the Voronoi cells, and an interfacial part due to the surface interactions between neighboring cells. Other than its heuristic nature, the main advantage of our geometrical approach lies in the fact that although it incorporates only nearest-neighbor effects, our calculations have a many-body character since both maximization of the packing density and minimization of the interfacial area are global rather than local problems.

We will now discuss the details of the two respective components of the lattice free energy for the two systems which are of interest to us, fuzzy and charged colloidal crystals. We emphasize that, due to the nature of the bulk energy as it relates to the space available to each colloidal particle's thermal movement as they arrange themselves in a lattice, it is only a function of lattice types and thus geometry. On the other hand, the different interaction types between colloidal particles are being encoded in the interfacial energy terms which make them system-specific.

5.2   Bulk Free Energy

To compute the bulk free energy of the system, we employ cellular free-volume theory [9, 10]. In this approximation, each particle is confined to a cage formed by its neighbors. The free volume available to each particle's center of mass is the volume of the Wigner-Seitz cell after a layer of thickness $\sigma/2$ (where σ is the hard-core diameter of particles) is peeled off from its faces. Since the volume of the Voronoi cell depends only on the symmetry of the lattice, the bulk free energy encodes information on the geometry of the system. The configuration integral of a one-component classical system of N particles confined to volume V is as follows:

$$Z = \frac{1}{\lambda^{3N} N!} \int_V \prod_{i=1}^{N} d\mathbf{r}_i \exp(-U(\mathbf{r_1}, \mathbf{r_2}, \ldots, \mathbf{r_N})/k_B T), \quad (5.1)$$

where the thermal de Broglie wavelength is $\lambda = \sqrt{h^2/2\pi m k_B T}$; the pairwise interaction is given by $u(\mathbf{r_i}, \mathbf{r_j}) = u(|\mathbf{r_i} - \mathbf{r_j}|)$, and the total energy is $U(\mathbf{r_1}, \mathbf{r_2}, \ldots, \mathbf{r_N}) = \frac{1}{2} \sum_{i,j=1}^{N} u(\mathbf{r_i}, \mathbf{r_j})$. Since each particle is assumed not to have any correlations with its neighbors, we have the following form for the partition function Z:

$$Z \approx \left(\lambda^{-3} \int_{V_0} d^3r \, \exp(-u_{\text{eff}}(r)/k_B T) \right)^N, \quad (5.2)$$

$$Z \approx \left(\lambda^{-3} \int_{V_0} d^3r \right)^N \approx \left(V_0/\lambda^3 \right)^N \quad (5.3)$$

where V_0 is the free volume of the cell, u_{eff} is the effective potential at each particle location (zero within the volume cell and infinite otherwise).

It turns out that the free volume theory yields good quantitative agreement with available numerical simulations at high densities in spite of its mean-field nature [6]. At lower densities, as long as the shear elastic constants are non-zero, we expect that an "Einstein-crystal" description of the phonon modes should be adequate. Indeed, for the systems in which we are interested, the appropriate moduli are all non-zero in the density regime we are probing [11]: therefore there should be no "soft modes" which might otherwise contribute strongly to collective effects.

In our work, we are mainly interested in three lattice candidates for the solid phase of fuzzy and charged systems; namely, the face-centered cubic (FCC), body-centered cubic (BCC), and β-tungsten (A15) lattices (Fig. 5.1). Survey of pertinent facts for these three lattices can be found in Section 4.4. The last lattice candidate (A15) has been observed in crystals of self-assembled micelles of some dendritic polymers [12]. This lattice belongs to the cubic system, and its unit cell includes eight sites which can be divided into three pairs of columnar sites and two interstitial sites. The columnar sites lie evenly spaced along the bisectors of the faces of the unit cell and can be thought of as forming three mutually perpendicular and interlocking columns. The interstitial sites fill out the space between the columns: one is at the center of the cell, and the other one is at the vertex. The A15 lattice possesses the Pm$\bar{3}$n space-group symmetry.

In the FCC lattice, the free volume has the shape of a rhombic dodecahedron just as the geometry of its Voronoi cell (Fig. 5.2). On the other hand, in the BCC lattice it remains an orthic tetrakaidecahedron only at rather low densities far below the freezing point. At higher densities, where the hard spheres form a solid phase, the square faces become absent rendering the free volume a regular octahedron. Since the A15 lattice has two types of lattice sites, its Voronoi cell correspondingly has two types of cell units and has the geometry of the Weaire-Phelan minimal surface.

The use of the free volume theory results in the following expression for the bulk free energy of the FCC or BCC lattice:

$$F_{\text{bulk}}^{X} = -k_B T \ln \left[\alpha^X \left(\beta^X n^{-1/3} - 1 \right)^3 \right], \qquad (5.4)$$

where $n = \rho \sigma^3$ is the reduced density and σ is the hard-core diameter of each colloidal particle. The coefficients $\alpha^{\text{FCC}} = 2^{5/2}$ and $\alpha^{\text{BCC}} = 2^2 3^{1/2}$ depend on the shape of the cells, whereas $\beta^{\text{FCC}} = 2^{1/6}$ and $\beta^{\text{BCC}} = 2^{-2/3} 3^{1/2}$ are determined by their size.

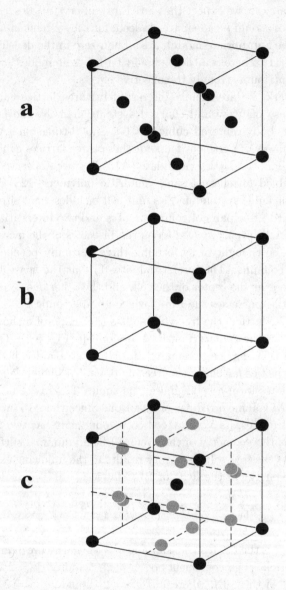

Fig. 5.1   Various lattices: (a) face-centered cubic, (b) body-centered cubic, and (c) A15 lattices.

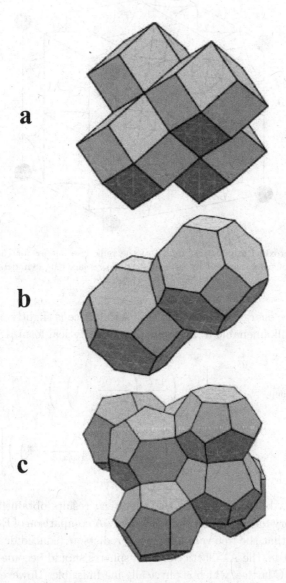

Fig. 5.2 Rhombic dodecahedron/FCC lattice (a), Kelvin's tetrakaidecahedron/BCC lattice (b), and Weaire-Phelan minimal surface/A15 lattice (c).

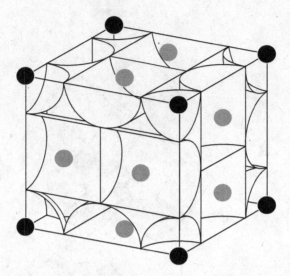

Fig. 5.3 Approximate analytical model of the bulk free energy for the A15 lattice: free volumes of columnar and interstitial sites are replaced by cylinders and spheres, respectively.

The bulk free energy expression for the A15 lattice is slightly more complicated but still amenable to an approximate analytical form [1, 2]:

$$F_{\text{bulk}}^{\text{A15}} = -k_B T \left[\frac{1}{4} \ln \left(\frac{4\pi S}{3} \left(\frac{\sqrt{5}}{2n^{1/3}} - 1 \right)^3 \right) \right.$$

$$\left. + \frac{3}{4} \ln \left(2\pi C \left(\frac{\sqrt{5}}{2n^{1/3}} - 1 \right)^2 \left(\frac{1}{n^{1/3}} - 1 \right) \right) \right]. \tag{5.5}$$

This formula best agrees with the numerical results obtained within the cellular theory for $S = 1.638$ and $C = 1.381$. A comparison of Eqs. (5.4) and (5.5) shows that the free volume theory predicts at reduced densities below about $n \approx 0.48$, the A15 lattice of hard spheres should become more stable than the FCC lattice, which is physically inadmissible. However, this is not essential since a pure hard-core system melts at reduced densities below $n \approx 1$. We note that this anomaly is not an artifact of the approximations behind Eq. (5.5); instead, it provides us with a conservative estimate for the range of validity of the cellular theory in this system (Fig. 5.4).

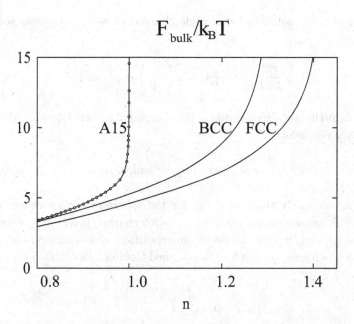

Fig. 5.4 Bulk free energies of hard-core particles arranged in FCC, BCC, and A15 lattice as calculated with the free-volume theory. Solid lines correspond to Eqs. (5.4) and (5.5), and circles are numerical results.

5.3 Interfacial Free Energy

5.3.1 *Charged Colloidal Crystals*

The interfacial free energy of the system, in contrast to the bulk free energy, incorporates the specific nature of interparticle interactions at the microscopic level. For charged systems, the interaction is the screened Coulomb potential. We neglect the effect of curvature and aim at arriving at the simplest expression capturing the essential physics of these systems. Assuming a system of 1:1 electrolytes, the behavior of the system is characterized by an electrostatic potential Ψ satisfying the linearized Poisson-Boltzmann equation:

$$\frac{d^2\Psi}{dx^2} = \frac{2e^2Z^2n_b}{\epsilon k_B T}\Psi \equiv \kappa^2\Phi, \tag{5.6}$$

where linearization is performed with respect to the dimensionless potential $\Phi \equiv \frac{e\Psi}{k_B T}$, n_b is salt concentration in bulk, ϵ is dielectric constant of solution, κ is the Debye decay length, and Z is the valence of the cationic species of interest. The solution for a single charged plate in electrolyte solution follows:

$$\frac{e\Psi}{k_B T} \equiv \Phi \sim 4\tanh\left(\frac{\Phi_s}{4}\right)\exp(-\kappa x), \qquad (5.7)$$

where Φ_s is the surface potential of the charged plate satisfying the following boundary condition:

$$q = 2(2\epsilon k_B T n_b)^{1/2}\sinh(\Phi_s), \qquad (5.8)$$

which is essentially the Gauss' law for the sphere of surface charge q. To obtain the interaction potential of two such charged plates at a distance d from each other, we use the linear superposition approximation and arrive at the following expression for the screened Coulomb interaction energy per unit area:

$$F_c = n_b k_B T \kappa^{-1}\Phi^2, \qquad (5.9)$$

$$F_c = 64 A_m k_B T n_b \kappa^{-1}\tanh^2\left(\frac{1}{4}\Psi_s\right)\exp\left(-\kappa d\right), \qquad (5.10)$$

where n_b is the bulk counterion number density, Ψ_s is the dimensionless surface potential of colloids, n is the colloid density, and the Debye screening length is $\kappa^{-1} = \sqrt{\epsilon\epsilon_0 k_B T/2e^2 Z^2 n_b}$, itself a function of these control variables (where $\epsilon\epsilon_0$ is the dielectric constant). Note that in obtaining the interfacial contribution to the free energy for charged colloids, we make use of the Debye-Hückel approximation, a Derjaguin-like approximation, as well as the linear-superposition approximation. We will show that, for the purpose of experimental comparison, our expression sufficiently captures the underlying physics of the system and yields predictions that compare reasonably with experimental data. It is conceivable that the level of accuracy should improve with a more exact, yet also necessarily more complicated, analytical expression of the charged interfacial free energy. For a different approach in summing up neighboring interaction between particles, we refer interested readers to a recent work in [26], wherein an explicit sum of electrostatic interaction between nearest neighbors is taken and the result demonstrated to be convergent in the two-dimensional case.

To relate the interfacial free energy in Eq. (5.10) back to the density of colloidal particles in the system, for the purpose of using it in conjunction with the bulk free energy to form the total free energy, we apply the fixed-volume geometric constraint, first introduced in Section 4.5,

$$A_m d = 2 \left(\frac{1}{n} - \frac{\pi}{6} \right) \sigma^3, \tag{5.11}$$

where $A_m = \gamma^x \sigma^2 n^{-2/3}$, which depends on the lattice type ($\gamma^{FCC} = 5.345, \gamma^{BCC} = 5.308, \gamma^{A15} = 5.288$) and is a dimensionless quantity characterizing the magnitude of the ratio of surface area to volume per cell. Again, this constraint comes about when we envision each colloidal particle in the lattice being enveloped by its respective Voronoi cell. Given the total volume of the system must be fixed at each density, so must be the sum of the volume of the hard cores plus the volume of the interstitial spacings. Using simple geometric considerations, we readily arrive at Eq. (5.11).

When we substitute Eq. (5.11) into the interfacial free-energy expression in Eq. (5.10), we relate the spacing d to the interfacial area A_m, which encodes geometric information on the lattice, as well as to the particle density n. Combining it with our expression for the bulk free energy, which is also a function of lattice type (geometry) and particle density, we now equip ourselves with the complete expression that should in principle encode all thermodynamical information on the system, and from which we can completely determine such issues as phase stability and computing all other related static quantities characterizing the system.

5.3.2 *Fuzzy Colloidal Crystals*

For fuzzy colloids, an argument based on Flory theory of the highly compressed polymer brushes [1, 2] with their excluded-volume repulsion between chains, yields the following interfacial free energy:

$$F_{\text{surf}} = \frac{\ell N_0 k_B T}{h} = \frac{2\ell N_0 k_B T}{d}, \tag{5.12}$$

where ℓ is a parameter with the dimension of length that determines the strength of repulsion, N_0 is the number of alkyl chains per micelle, and h, the thickness of the corona, is half the average thickness of the interdigitated matrix of the chains, d, as shown in Fig. 5.5. In both systems, maximization of surface spacing d implies the minimization of the interfacial area A_m. From Eq. (4.24), the density of particles enter directly into the energies of the system through the minimal-area constraint. However, the functional dependence on the particle density differs between them: for the charged

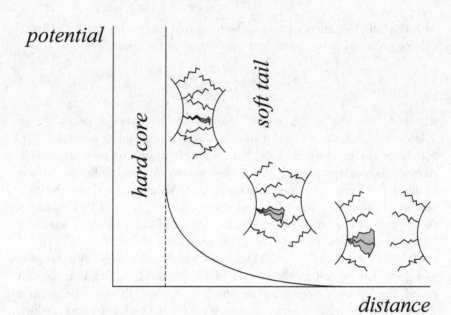

Fig. 5.5   Interactions between fuzzy colloids. For colloidal particles of the fuzzy kind, there are two main types of interactions: hardcore interaction stemming from the non-penetrability between the particle cores; soft-tail interactions between neighboring particles that increase with decreasing interparticle distances.

colloids, we have a short-range exponential interaction while for the fuzzy colloids we have an algebraic interaction, valid at ranges smaller than the thickness of an uncompressed coronal brush.

5.4   Conclusion

The bulk and interfacial free energies together express the thermodynamics of colloidal systems in a geometrical language. The formalism directly establishes relations between microscopic parameters governing the dynamics of the colloidal particles and the stability of various macroscopic phases with respect to temperature [3, 2]. The frustration arising from the system's inability to maximize packing density and minimize interparticle repulsion simultaneously in fuzzy and charged colloidal crystals are responsible for the variety of non-close packed structures observed experimentally. The theoretical investigation of these two determining factors is deeply connected to the mathematical problem of Kelvin and Kelper conjecture. One can conceivably adapt Eqs. (5.4), (5.5), (5.10), and (5.12) to other kinds of lattice geometry, and the foam analogy can correspondingly be applied to studying the various other solid phases of colloidal systems. However, we will work mainly with the FCC, BCC, and the A15 lattices in this work.

References

[1] P. Ziherl and R. D. Kamien, *Phys. Rev. Lett.* **85**, 3528 (2000).

[2] P. Ziherl and R. D. Kamien, *J. Phys. Chem. B* **105**, 10147 (2001).

[3] W. Kung, P. Ziherl and R. D. Kamien, *Phys. Rev. E* **65**, 050401(R) (2002).

[4] R. L. Whetten, M. N. Shafigullin, J. T. Khoury, T. G. Schaaff, I. Vezmar, M. M. Alvarez and A. Wilkinson, *Acc. Chem. Res.* **32**, 397 (1999).

[5] E. B. Sirota, H. D. Ou-Yang, S. K. Sinha, P. M. Chaikin, J. D. Axe and Y. Fujii, *Phys. Rev. Lett.* **62**, 1524 (1989).

[6] W. A. Curtin and K. Runge, *Phys. Rev. A* **35**, 4755 (1987).

[7] A. R. Denton and N. W. Ashcroft, *Phys. Rev. A* **39**, 4701 (1989).

[8] G. A. McConnell and A. P. Gast, *Phys. Rev. E* **54**, 5448 (1996).

[9] T. L. Hill, *Statistical Mechanics*, McGraw-Hill, New York, 1956.

[10] J. A. Barker, *Lattice Theory of the Liquid State*, Pergamon Press, Oxford, 1963.

[11] We thank M. O. Robbins for discussions on this point.

[12] V. S. K. Balagurusamy, G. Ungar, V. Percec and G. Johansson, *J. Am. Chem. Soc.*, **119**, 1539 (1997).

[13] T. C. Hales, *Notices Amer. Math. Soc.*, **47**, 440 (2000).

[14] W. Kohn and L. J. Sham, *Phys. Rev.*, **140**, 1133A 1965.

[15] J. Z. Wu, *AIChE J.* **52**, 1169 (2006).

[16] M. Rex, H. H. Wensink, and H. Lowen, *Phys. Rev. E* **76**, 021403 (2007).

[17] J. Kleis, B. L. Lundqvist, D. C. Langreth, and E. Schroder, *Phys. Rev. B* **76**, 100201(R) (2007).

[18] P. C. Hiemenz and R. Rajagopalan, *Principles of Colloid and Surface Chemistry*, CRC, New York, 1997.

[19] B. Mennucci and R. Cammi, *Continuum Solvation Models in Chemical Physics: From Theory to Applications*, Wiley Press, New York, 2008; F. Hirata, *Molecular Theory of Solvation*, Springer Verlag, New York, 2003.

[20] P. G. de Gennes and J. Prost, *The Physics of Liquid Crystals*, Oxford University Press, Oxford, 1995; P. J. Collings, *Liquid Crystals: Nature's Delicate Phase of Matter*, Princeton University Press, Princeton, 2001.

[21] W. B. Russel, D. A. Saville, and W. R. Schowalter, *Colloidal Dispersions*, Cambridge University Press, Cambridge, 1992; R. J. Hunter, *Foundations of Colloid Science*, Oxford University Press, Oxford, 2001.

[22] T. V. Ramakrishnan and M. Yussouff, *Phys. Rev. B* **19**, 2775 (1979).

[23] W. A. Curtin and N. W. Ashcroft, *Phys. Rev. A* **32**, 2909 (1985).

[24] P. M. Chaikin and T. C. Lubensky, *Principles of Condensed Matter Physics*, Cambridge University Press, New York, 2000.

[25] W. D. Callister Jr., *Materials Science and Engineering: An Introduction*, John Wiley & Sons, New Jersey, 2003.

[26] W. Kung and M. Olvera de la Cruz, *J. Chem. Phys.* **127**, 244907 (2007).

Chapter 6

Phases of Charged Colloidal Crystals

6.1 Introduction

In this chapter, we apply the minimal-area conjecture to systems of charged colloidal suspensions. Previous works have addressed colloidal particles characterized by hard cores with fluffy coronas, and the minimal-area conjecture has been proposed as a mean-level formalism to account for the degrees of freedom associated with these coronas. Whereas these systems are modeled by the inter-particle potential of simple hard cores dressed with a short-range repulsion, the charged systems considered in the present chapter interact mainly via an infinite-range, screened Coulomb potential. In this work, we study the solid-phase transitions of charged-colloidal suspension as a function of particle volume fraction and ionic strength. Using the Debye-Huckel approximation and introducing the minimal-area constraint, we derive the main features of a phase diagram for the charged system studied in a previous experiment, with the surface potential of the colloidal spheres being the one adjustable parameter in the data fit. We further calculate the shear modulus [c.f. Chapter 7] of the system using our model and find agreement with the available data.

Charged colloids have been the subject of intense study both experimentally and theoretically. In the laboratory, systems of charged suspensions can be studied readily with optical techniques and manipulated easily with chemical means. Their rich chemistry leads to many industrial applications ranging from uses in coating materials, in ceramic precursors, and in designing and manufacturing biological macromolecules. Theoretically, these systems interact via an infinite-range screened Coulomb potential and thus epitomize one of the few classes of possible interactions in condensed-matter systems. Thus, the motivation to a better understanding of the phase behavior of these charged systems can hardly be over-emphasized.

Previously, many experimental studies provided a wealth of data with regard to the stability of various phases, notably the disordered phase, the face-centered cubic phase (FCC) and the body-centered cubic phase (BCC). These systems consisted typically of aqueous suspension of uniform, charged- polystyrene spheres with variable salt concentration, the latter being a control parameter of the degree of screening of the underlying Coulomb interaction. For the ordered phases, it is often the case in which the interparticle distance is large and thus completely amenable to probing by scattering of visible light [1]. By such means, observed are the general trends in which the order-disorder transition occurs at low salt concentrations and in which the structural FCC-BCC transition occurs at high volume fractions. Numerous molecular dynamics (MD) simulations [2] based on the Yukawa potential also supported these experimental findings. Though it is encouraging that such qualitative agreement exists in the phase characterization of these systems, one is hard-pressed for similar quantitative convergence between existing experimental data and simulation results. Further complicating the matter is the lack of analytical calculations from which either experimental or simulation findings can be definitively derived. This difficulty thus strongly necessitates formulation of theoretical models that are not only computationally simple enough to yield insightful solutions but also general enough to have great applicability to a range of diverse systems.

In what follows, we apply the framework of foam analogy to the case of charged colloids. We construct a model containing one free parameter of the surface potential: it couples the bulk free energy of the charged hard cores to the free energy of their screened Coulomb interaction via the minimal-area constraint. For a given value of the surface potential, we derive the phase diagram obtained in a previous experimental study. In the next chapter, following simple analysis using elasticity theory, we will also derive a shear modulus comparable to the quoted experimental value. In short, we demonstrate the plausibility of understanding the various experimental findings in terms of the geometrical foam model already established for another class of systems.

6.2 Phase Transitions of Charged Colloids

Among the earliest experimental studies of systems of charged colloidal suspensions was that of Vanderhoof et al. [3] They first noticed the iridescence of ordered de-ionized lattices with large interparticle spacing of up to five times the particle diameter. The conclusion was that the form of interparticle interaction must be electromagnetic in origin, since at such interparticle distance the attractive Van der Waals force would be much weaker. Hachisu et al. [4] collected the most comprehensive data regarding phase transitions of aqueous polystyrene suspensions. However, surface charges of these systems were not reported, and the structural transformations were not investigated at low ionic strengths. Nevertheless, the order-disorder as well as the FCC-BCC structural transitions were found in agreement with those in simulation studies for hard spheres. In regard to the characterization of such systems, scattering of visible lights was the most often used. Hiltner and Krieger [5] determined the lattice constant using Bragg reflection techniques. Other studies concentrated on such properties as the bulk modulus [6], shear deformations [7], and time-dependent ordering processes of the ordered phase [8], based on analysis of the Kossel lines [9] created by light rays being backscattered at the Bragg angle from imperfections in the crystal. As mentioned in the introduction, the order-disorder transition was observed at low ionic strengths and the FCC-BCC transition at low volume fractions. Specifically, it is believed that FCC structure generally occurs at volume fraction $\phi > 0.03$ and that BCC structure at $\phi < 0.02$ [10].

To understand the behavior of phase transitions of charged colloids, several theoretical models were proposed. Both the Derjaguin approximation and the linear superposition approximation [10] were used in modeling the interaction potential between charged spheres. For many cases in which the inter-particle distance is much larger than the radius of each spherical colloidal particle, the linear-superposition approximation sufficiently captures the essence of screened Coulomb interaction. In most models of the order-disorder transition, the fluid phase is modeled by hard spheres whose effective radii can be calculated by the Barker-Henderson perturbation. This approach works well for high ionic strengths since the screening of Coulomb potential is strong and the hard sphere approximation becomes highly accurate. In fact, Van Megen and Snook [11] showed excellent agreement between the order-disorder transition predicted by Barker-Henderson perturbation [12] and Monte Carlo simulation results at moderate ionic strengths.

Similarly, proposals were made to predict quantitatively behavior of the FCC-BCC transition. Beunen and White [13] determined the potential energies of the two structures based on the assumption of only nearest-neighbor interaction. Their results were similar to those obtained by Silva and Mokross [14] who looked at the internal energy difference between the two phases, as well as to those by Chaikin et al. [15] who instead investigated the corresponding free energy difference. The main underlying assumption in the aforementioned studies is that of colloidal particles behaving as point-like particles. Hone et al. [16] also made the same assumption in his study along with the use of the Lindemann rule to estimate the melting curve. This, however, led to the prediction of reentrant BCC-FCC-BCC transition upon increasing the salt concentration that was not observed in experiments. In yet another similar study by Shih et al. [17], they actually accounted for the finite size of colloidal particles in their variational-principle formalism, and as a result such reentrant transition was not predicted. Instead, the FCC structure was found to be stable at higher volume fractions and the BCC structure at lower volume fractions in their highly charged system. Note that the previous models all suffered from the lack of the need to incorporate the exact nature of the solid phase in their calculations; in perturbation theory one simply treats solids analytically the same as fluids but with higher volume fractions. Fortunately, progress was made towards more realistic calculations such as those found in the MD simulation by Kremer, Robbins, and Gest [2] that include soft repulsions for the solid phase.

In our foam model, we connect the main findings from the above works to aspects of the Kelvin's problem [18]; the Kelvin's problem asks for a regular partition of space into cells of equal volume each having the smallest area. This connection is achieved by coupling the free energy expression of the ordered structures to the minimal-area constraint, previously proposed as an intuitive explanation for an abundance of non-close packed crystals observed in another class of system: non-charged colloids comprising hard cores and fluffy corona [19, 20]. In effect, rather than mapping the problem to the limit of infinite curvature where colloidal particles are approximated by points, we consider the opposing limit of zero curvature describable in the language of soap froths of the Kelvin's problem. In what follows, we shall discuss the previously established minimal-area conjecture in the present context of charged systems and examine the consequences.

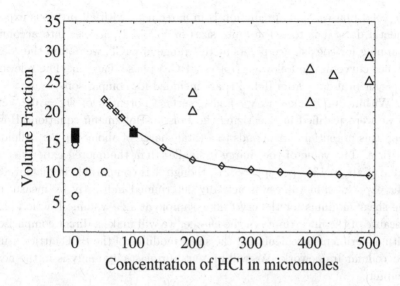

Fig. 6.1 Theoretical FCC-BCC coexistence curve as a function of volume fraction ϕ and electrolyte concentration (HCl). The diamonds are the theoretical predictions, while the other points come from the data of Ref. [21]. Solid squares are coexistence points, open triangles are FCC, and open circles are BCC.

6.3 Foam Analogy and Charged Colloids

Focusing on the structural phase transition from BCC to FCC, we rely on the X-ray scattering study of charged polystyrene spheres by Sirota et al. [21]. That experiment studied the phase diagram of polyballs in a 0.9-methanol–0.1-water suspension. The thermodynamic behavior of this system was recorded as a function of volume fraction ϕ and salt concentration c_{HCl}. In order to determine the surface potential, we choose an FCC-BCC coexistence point in the experimental data, and then equate the corresponding FCC and BCC free energies to solve for Ψ_s. Through Gauss' law, we calculate the charge per colloid $q = 2A_M\sqrt{2\epsilon k_B T n_b}\,\sinh\left(\frac{1}{2}\Psi_s\right)$. Finally, using this value of the charge, we find the density n at which $F^{FCC}/k_B T$ and $F^{BCC}/k_B T$ are equal for each salt concentration n_b. Due to the limited data, there is only one point at non-zero salt concentration that is in the coexistence region. We thus obtain the coexistence curve in $(\phi\text{-}n_b)$ phase diagram (Fig. 6.1). In the range of salt concentrations considered, from [HCl] = 10 μmol to 500 μmol, the free energies for both phases are convex in their support. Thus there are no isostructural transitions in the solid phases. This prediction is in agreement with all previous experimental data that took the finite sizes of colloidal particles into account. Scanning for coexistence points at this range of [HCl], we obtain the coexistence curve in the following ($[\phi]$ vs. [HCl]) phase diagram. The relevant experimental data from Ref. [21] are included for comparison.

Within our framework we find a surface potential of $\Psi_s = 0.2$, and so we were justified in linearizing the Poisson-Boltzmann equation. However, this potential corresponds to a total charge of about $48e$ per colloidal particle. The value of the charge is smaller than the quoted experimental value of about $135e$ per sphere [21], though it is comparable in magnitude. The experimental value was actually determined indirectly by measuring the shear modulus for the crystalline sample at 12% volume fraction [22]. Because of the uncertainty in the charge, we will make a direct comparison with experiment by calculating the shear modulus of the BCC lattice using our cellular framework. We will present our elasticity analysis in the next chapter.

6.4 Conclusion

We have established a geometrical framework for understanding the structural properties of charged colloids. We have derived the phase diagram of the system with only one adjustable parameter, the surface charge. The balance between the drive of the system's entropy for maximal packing fraction and the need of its screened repulsion for minimal interfacial area accounts for most of the properties found in experiments. Its elucidation can serve as an intuitive guide to the engineering of all such charged systems.

References

[1] Y. Monovoukas and A. P. Gast, *J. Col. Int. Sci.* **128**, 533 (1989).

[2] M. O. Robbins, K. Kremer and G. S. Grest, *J. Chem. Phys.* **88**, 3286 (1988).

[3] J. W. Vanderhoff, H. J. van den Hul, R. J. M. Tausk and J. T. G. Overbeek, *Clean Surfaces: Their Preparation and Characterization for Interfacial Studies*, Decker, New York, 1970.

[4] S. Hachisu,Y. Kobayashi and A. Kose, *J. Col. Int. Sci.* **342**, 42 (1973).

[5] A. P. Hiltner and I. M. Kreiger, *J. Phys. Chem.* **73**, 2386 (1969).

[6] R. J. Carlson and S. A. Asher, *Appl. Spectrosc.* **38**, 297 (1984).

[7] B. J. Ackerson and N.'A. Clark, *Phys. Rev. Lett.* **46**, 123 (1981).

[8] T. Yoshiyama, *Polymer* **27**, 828 (1986).

[9] E. Dubois-Violette, P. Pieranski, F. Rothen and L. Strzelecki, *J. Physique* **41**, 369 (1980).

[10] W. B. Russel, D. A. Saville and W. R. Schowalter, *Colloidal Dispersions* Cambridge University Press, New York, 1989.

[11] W. Van Megen and I. Snook, *J. Col. Int. Sci.* **57**, 40 (1976).

[12] A. P. Gast, C. K. Hall and W. B. Russel, *Faraday Discuss. Chem. Soc.* **76**, 189 (1983).

[13] J. A. Beunen and L. R. White, *Col. Surf.* **3**, 317 (1981).

[14] J. M. Sílva and B. J. Mokross, *Phys. Rev. B* **21**, 2972 (1980).

[15] P. M. Chaikin, P. Pincus and S. Alexander, *J. Col. Int. Sci.* **89**, 555 (1982).

[16] D. Hone, S. Alexander, P.M. Chaikin and P. Pincus, *J. Chem. Phys.* **79**, 1474 (1983).

[17] W. Y. Shih, I. A. Aksay and R. Kikuchi, *J. Chem. Phys.* **86**, 5127 (1987).

[18] W. Thomson, *Phil. Mag.* **24**, 503 (1887).

[19] P. Ziherl and R. D. Kamien, *Phys. Rev. Lett.* **85**, 3528 (2000).

[20] P. Ziherl and R. D. Kamien, *J. Chem. Phys. B* **105**, 10147 (2001).

[21] E. B. Sirota, et al., *Phys. Rev. Lett.* **62**, 1524 (1989).

[22] J. F. Joanny, *J. Col. Int. Sci.* **71**, 622 (1979).

Chapter 7

Elasticity of Colloidal Crystals

7.1 Introduction

In Sirota's experiment [1], the authors determined the average shear modulus for the crystalline sample at 12% volume fraction via a scaling argument by Joanny [2]. The average shear for an isotropic sample is related to the density ρ and the interaction potential $V(a)$ as a function of interparticle spacing a, as follows:

$$\mu = \frac{4}{9}\rho V(a)(\kappa a)^2;$$

$$V(a) = \frac{Z^2 e^2}{\epsilon a}\exp(-\kappa a) \qquad (7.1)$$

Therefore, the effective surface charge of each colloidal particle in the system can be found from Eq. (7.1) with the appropriate substitution of the experimental value of the isotropic shear modulus $\mu\,(= 17\,\mathrm{N/m}^2)$, density $n\,(= 0.23)$, the interparticle spacing $a\,(= 2.94 \times 10^7\mathrm{m})$, the dielectric constant of the water/methanol solution $\epsilon\,(= 38\epsilon_0)$ for $\lambda = \kappa a\,(\simeq 4.64)$. The effective charge Z was determined to be approximately $135e$ per sphere, as quoted earlier.

While the surface potential of these charged colloidal particles are typically treated as a fitting parameter in experiments, there are no definitive way of measuring the surface charge of these particles experimentally. Our model predicts a somewhat small surface charge for the colloidal spheres of the system. While its magnitude is of an unusual order, it is quite common for the effective charge to be different from and in fact smaller than the net or titratable charge.

To calculate the renormalized charge Z^* and acid concentration n^*_{HCl}, Alexander et al. [3] employed a Wigner-Seitz approximation similar in principle to our way of determining the bulk free energies of the various lattices. To ensure analytical tractability of the solution, a sphere of the same volume replaces the true Wigner-Seitz cell of the system under study. Values of Z^* and n^*_{HCl} are then determined from matching the solution of the corresponding Poisson-Boltzmann equation and its first three derivatives to the effective Yukawa potential at the cell boundary. It is argued by Alexander et al. [3] that the renormalized salt density is insignificant and can be ignored in most circumstances. The rationale behind this method is that in interstitial regions, which occupy a large fraction of the system, the potential Φ does not have great variations and can suitably be determined from the linearization of the corresponding Poisson-Boltzmann equation. Moreover, these interstitial regions predominately determine the screening rate of the system. The only region where the above approximation fails to apply is that near the colloidal particles themselves. Both the potential Φ and the charge densities vary rapidly in these regions. As a result, the counterions are more tightly constrained to locations near the surfaces of these colloidal particles.

Therefore, one can essentially interpret these ions as part of an effective sphere with a reduced charge. The applicability of linearization of the Poisson-Boltzmann equation to charged colloidal systems could now be greatly enhanced providing that proper determinations of the renormalized effective charge and of the salt concentration are made. Unfortunately, both the effective charge and the ion concentration cannot be measured directly in experiments, making the confirmation of their values extremely difficult. Based on the considerations thus far, the prediction of the smaller surface charge density from our model is certainly plausible.

Another experimental complication for the present case of charged colloids is that the fine-tuning of system properties is less than straightforward. One can only modify the salt concentrations to adjust the screening length indirectly. Moreover, one cannot determine precisely and thus control the bare charge of the particles, since ionization of the surface acidic groups is always incomplete.

The usual method of determining such bare charges is through titration. Other methods include measuring solution conductivity, force constants, bulk moduli, and shear moduli. Bulk moduli can be measured by detecting density changes in the presence of gravity, and shear moduli can be studied readily using resonance methods. Recently, the microscopic measurement of the pair-interaction potential of charge-stabilized colloid was performed by Crocker and Grier [4]. Their method was to study the motion of random pairs of particles, governed mathematically by the discretized version of the Smoluchowski equation whose initial positions were fixed physically using optical tweezers and subsequent motions captured on film. The experimental problem of determining the interaction potential can then be reduced to that of characterizing the discrete transition probability matrix. Their finding of the effective charge for their particular system did confirm qualitatively the charge renormalization calculations proposed by Alexander et al. [3] and lent support to the notion of a generally smaller effective surface charge on the spheres.

Nevertheless, instead of relying on such indirect methods in establishing connections between experimental data and theoretical framework via concepts such as potential interaction and surface charge density, we extend our foam analogy construction to a direct computation of the elastic constants, from which the bulk modulus as well shear modulus can be derived based upon first principles.

7.2   Foam Analogy and Cubic Elastic Constants

In general, the elastic free energy for a solid is quadratic in the symmetrized strains as follows [5]:

$$F_{el} = \frac{1}{2} \int d^d x \, K_{ijmn} \, u_{ij} u_{mn} \,. \tag{7.2}$$

By definition, since $u_{ij} = u_{[ij]}$, the elasticity tensor is invariant under the interchange of the first two indices or the last two indices:

$$K_{ijmn} = K_{[ij][mn]} \,. \tag{7.3}$$

This symmetry reduces the number of independent components to 36. Furthermore, the elastic constant tensor is also invariant under the interchange of the first pair of indices with the second pair, as follows:

$$K_{[ij][mn]} = K_{[mn][ij]} \,. \tag{7.4}$$

Therefore, for any triclinic crystalline system, there are only 21 independent elastic constants.

For a crystal with cubic symmetry, however, there exists additional symmetries such that the total number of independent components of the elastic constant tensor can be further reduced. In particular, there exists a four-fold rotational symmetry (C_4) about the crystal's body axis so the physical properties do not change upon a 90°-rotation, i.e. $(x \leftrightarrow y)$. There is one additional mirror symmetry σ such that upon reflection about any plane perpendicular to one of the axes, the crystal retains the same physical property, i.e. $(x \leftrightarrow -x)$. As such, there are only three independent components as follows:

$$K_{xxxx} \equiv K_{11}$$
$$K_{xxyy} \equiv K_{12} \tag{7.5}$$
$$K_{xyxy} \equiv K_{44}$$

and the elastic free energy of a crystal with cubic symmetry can now be written in terms of these three distinct elastic constants [5]:

$$F_{\text{cubic}} = \frac{1}{2} \int d^3x \Big[K_{11} \left(u_{xx}^2 + u_{yy}^2 + u_{zz}^2 \right) + K_{12} \left(u_{xx}u_{yy} + u_{xx}u_{zz} + u_{yy}u_{zz} \right)$$
$$+ 2K_{44} \left(u_{xy}^2 + u_{xz}^2 + u_{yz}^2 \right) \Big]. \tag{7.6}$$

The bulk modulus $K = \frac{1}{3}(K_{11} + K_{12})$ is an isotropic quantity. The shear modulus, on the other hand, generally depends on the direction of the applied strain and ranges from deformations along the four-fold axis (elongational shear), $\mu = K_{11} - \frac{1}{2}K_{12}$, to deformations along the face diagonal (simple shear), K_{44}. To determine the bulk modulus as well as the three elastic constants, we calculate the changes both in the free volume and in surface area associated with the deformations of the lattice in question (Fig. 7.1).

The bulk modulus can be expressed in terms of the total free energy of the system via:

$$K = V \left(\frac{\partial^2 F}{\partial V^2} \right)_T = \frac{1}{3} \left(K_{11} + K_{12} \right), \tag{7.7}$$

where V is the volume of the crystal. To compute the three elastic constants K_{11}, K_{12}, and K_{44} for the FCC, BCC, and A15 lattices in our study, we need two more measurements which we attain from the elongational and simple shear modes.

The elongational shear mode is parametrized by the coordinate transformation: $(x, y, z) \to (x/\sqrt{1+q}, y/\sqrt{1+q}, (1+q)z)$ where $q \ll 1$, induces a

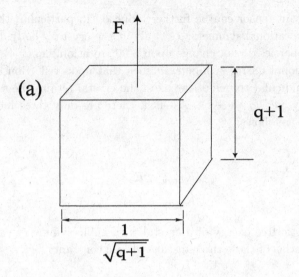

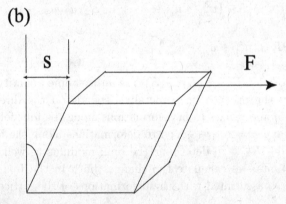

Fig. 7.1   The two different types of deformations: (a) elongational shear mode; shear deformation along the four-fold axis (b) simple shear mode; shear deformation along the face diagonal. In both cases, we only consider deformations which preserve the total volume of the unit cell. The elongational shear mode is parametrized by q, while the simple shear mode is parametrized by s.

corresponding strain $(u_{xx}, u_{yy}, u_{zz}) = (-q/2, -q/2, q)$. Upon substitution into Eq. (7.6), we find:

$$F_{\text{elong}}(q) = \frac{3V}{4}\left(K_{11} - \frac{1}{2}K_{12}\right)q^2. \qquad (7.8)$$

The volume V simply results from the integration over the free cellular volume in the definition of the elastic free energy. Similarly, the simple

F(q)

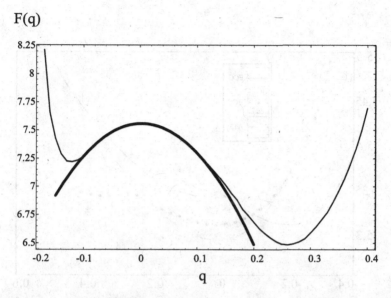

Fig. 7.2 The bulk free energy of the BCC lattice as a function of the elongational shear at $n = 1.1$. The thick line represents the quadratic fit to the curve at $q = 0$ which determines the bulk part of the shear modulus. The minimum at $q \approx 0.26$ corresponds to the FCC lattice.

shear mode, parametrized by the coordinate transformation $(x, y, z) \rightarrow (x, y, sx + z)$, induces the corresponding strain $u_{xz} = s/2$; all other components of the strain tensor vanish. Analogously, we have:

$$F_{\text{simple}}(s) = \frac{V}{4} K_{44} s^2. \tag{7.9}$$

The left-hand side of Eqs. (7.8) and (7.9) can be found within our framework. There are two contributions from the bulk and interfacial parts: we now find their dependence on the deformation parameters q and s. For example, we consider the elastic energy of the elongational shear of the BCC lattice in Fig. 7.2. Note that the magnitude of these negative contributions to the bulk modulus increases with increasing density.

The bulk elastic energies can be found via a simple numerical scheme by deforming the unit cell and calculating the resulting free volume of the colloidal particle. At this point, we note that this approach reproduces the well-known shear instability of the BCC lattice of hard spheres. Shear along a four-fold axis, called the Bain strain, reduces the bulk free energy and leads eventually to an FCC lattice via a continuous sequence of body-centered tetragonal (BCT) lattices. This is illustrated in Fig. 7.2 where the

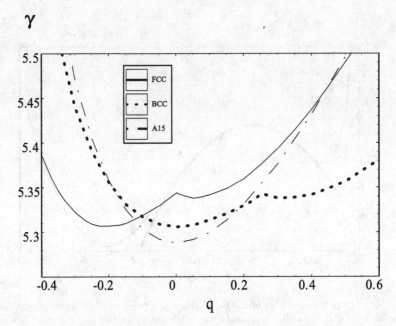

Fig. 7.3   The surface area parameter γ as function of dimensionless parameter q for the elongational shear mode. The kinks in the curves indicate a change in the topology of the unit cell.

bulk free energy of a BCC lattice is plotted as a function of the elongational shear parameter q. At $q = 0$, the unit cell is cubic and the free energy reaches a local maximum. The absolute minimum, which is located at $q = 2^{1/3} - 1 \approx 0.26$ at all densities, corresponds exactly to the FCC lattice. The position of the local minimum at $q < 0$, which describes a BCT lattice, depends on density. Its relative depth with respect to the BCC lattice increases with density. The modulus of the curvature near the BCC point increases with increasing reduced density n.

The computation of the interfacial elastic free energy was facilitated by Surface Evolver [6]. Referring to the expressions for the interfacial free energy, we can consider the change in γ with respect to the deformation parameters q and s. This function now encodes the change in surface area of each Voronoi cell under the elongational and simple shear modes, respectively. The numerical result for the elongational shear mode is shown in Fig. 7.3.

The stability of any particular lattice depends on both the bulk and surface free energy terms. In the BCC lattice, the bulk term favors the shear deformation whereas the surface term does not. It is their relative magnitude that determines whether the lattice is stable, and that depends on density. Since the bulk free energy diverges at close packing, we expect that the BCC lattice should always become unstable at high enough densities. In the following, we only plot the elastic constants for each lattice in the physically relevant range of parameters where the lattice is stable.

We fit the curves shown in Fig. 7.3 with a quadratic to extract a term proportional to the square of the strain parameter. Upon substitution into Eqs. (5.10) and (5.12), we obtain the interfacial contribution to the elastic energy for both the charged and fuzzy colloidal systems, respectively. We repeat the procedure for the set of bulk elastic energy curves (Fig. 7.2) and the analogous data set for the simple shear mode. The final step requires equating the continuum energy change from Eq. (7.6) with the calculated energy change and solving numerically for the coefficients in Eqs. (7.7), (7.8), and (7.9).

7.3 Elasticity of Charged Colloidal Crystals

To have a direct comparison with the experimentally measured isotropic shear modulus quotes by Sirota et al. we first compute the three cubic elastic constants K_{11}, K_{12}, K_{44}, and the bulk modulus K for the case of charged colloids in a BCC lattice. In particular, we find that $K_{11} \approx 17.1$ N/m^2, $K_{12} \approx -1.2$ N/m^3, and $K_{44} \approx 10.3$ N/m^2 for a 12% volume-fraction ($n = 0.23$) BCC sample. Thus the shear modulus ranges from 10.3 to 17.7 N/m^2. Again this is the same order of magnitude as the measured, isotropic value of 17 N/m^2 [1], which is a complicated combination of the two shear moduli. We have calculated these moduli using an HCl concentration of 50 μmol, as this is in the middle of the reported BCC regime [1].

In order to compare with MD simulations [7], we tabulate (Table 7.1) values of the bulk modulus and the three elastic constants as a function of $\lambda = \kappa a$ (in the experimental study, $\lambda \approx 4.64$). However, unlike point particles with Yukawa interactions, our model does not scale simply with this parameter. To vary λ, we keep the colloidal density fixed as well as the total charge on each colloid. Varying the salt concentration changes both κ and the surface potential Ψ_s. With these new values we can recalculate the elastic constants. For completeness, we tabulate the values of the bulk modulus K and elastic constants of the BCC structure at $\phi = 0.12$ for

Table 7.1. Calculated values for the bulk modulus K and elastic constants of the BCC structure at several values of $\lambda = \kappa a$, where a is the average interparticle spacing, for $n = 0.23$ (first number) and $n = 0.50$ (second number). All values are in units of N/m^2.

λ	K	K_{11}	K_{12}	K_{44}
1	−26.1 / −7.59	1.53 / 5.13	−79.8 / −27.8	54.4 / 28.4
2	−7.54 / 13.8	12.0 / 22.3	−35.2 / 19.3	29.1 / 15.6
3	0.12 / 22.3	16.5 / 29.0	−16.1 / 37.8	18.3 / 10.6
4	3.66 / 26.2	17.2 / 31.7	−6.18 / 46.7	12.7 / 7.95
5	5.37 / 28.0	16.5 / 32.7	−0.37 / 51.2	9.40 / 6.40
6	6.16 / 28.6	15.2 / 32.6	3.33 / 53.3	7.30 / 5.43
7	6.46 / 28.7	11.4 / 31.9	5.67 / 54.2	5.90 / 4.70
8	6.49 / 28.5	9.13 / 31.2	7.37 / 54.3	5.00 / 4.30

different values of $\lambda = \kappa a$ (where a is the mean separation of spheres; at $n = 0.23$, $a \approx 140$ nm).

In short, the complex nature of charged systems in solution generally renders them to be "messy" systems, since clean measurements of experimental quantities are difficult to obtain under the most ideal conditions. Neither the effective charge nor the local ion concentration can be measured directly in experiments, making the precise confirmation of theory for these charged systems difficult. Because our model allows the simultaneous prediction of the phase boundary and the shear moduli, we can avoid these uncertainties.

There are, of course, additional interactions that we have neglected. The van der Waals attraction is much weaker than the screened Coulomb potential at the interparticle spacing of $a \approx 140$ nm. We expect that correlation effects such as overcharging should have minimal consequence in a system of monovalent salt ions of NaCl [8]. Dispersion forces are important at ion concentrations greater than 100 μmol [9]. However, this effect occurs at distance scales on the order of 10 μm, and we do not expect this to change our model.

Finally, we note that in the calculation of the bulk free energy, we employed free-volume theory even in the relatively low-particle density regime — a regime in which pure hard-core interactions would predict a fluid phase. However, the screened Coulomb potential stabilizes the lattice structures and, in turn, lowers the melting density of the system. This fact

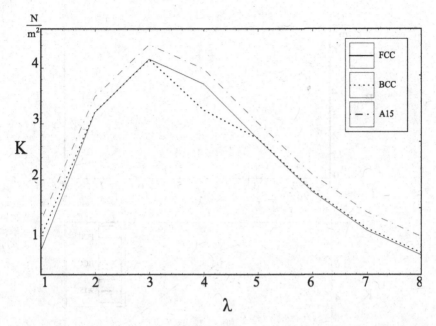

Fig. 7.4 The bulk modulus K for the FCC, BCC, and A15 lattices of a charged colloidal system as a function of $\lambda = \kappa a$, where a is the average interparticle spacing. The calculations are done at density $n = 0.9$ and at dimensionless surface potential $\Phi_s = 0.4$. The maximum at $\lambda = 3$ is spurious and signals the breakdown of our approximation (see text).

manifests itself in the moduli — our model predicts shear moduli on the order similar to the corresponding order of magnitude for the bulk moduli in the experimental regime of interest. We therefore believe that there are no soft modes which would allow for large deformations of the lattice. One could characterize the system with an effective density of higher magnitude, determined by the Barker-Henderson effective diameter of particles [10]. A scheme like this would introduce another parameter into our model, which would divide the screened-Coulomb potential into a "hard" part and a "soft" part.

We can easily extend our calculations to the FCC and A15 lattices (or any other lattices of interest). In each case, we examine the behavior of the bulk modulus as well as the three elastic constants K_{11}, K_{12}, and K_{44} for the three lattices with respect to both varying density and varying effective "screening" length.

In what follows, the surface potential and the effective diameter of the colloidal particles are chosen to be $\Phi_s = 0.4$ and $\sigma = 910$ nm, respectively.

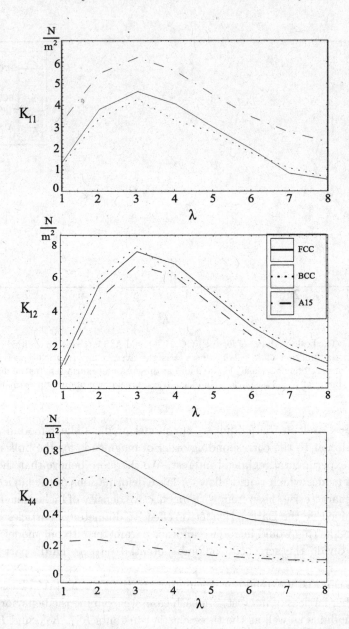

Fig. 7.5 The elastic constants K_{11}, K_{12}, and K_{44} for the FCC, BCC, and A15 lattices of a charged colloidal system as a function of $\lambda = \kappa a$, at density $n = 0.9$ and at dimensionless surface potential $\Phi_s = 0.4$.

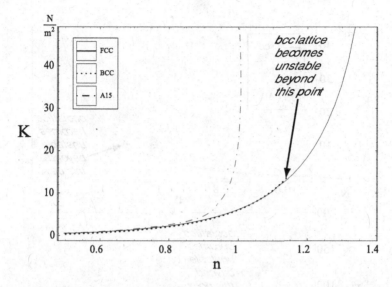

Fig. 7.6 The bulk modulus K for the FCC, BCC, and A15 lattices of a charged colloidal system as a function of density n, at fixed screening length $\lambda = 4$ and at surface potential $\Phi_s = 0.4$. The bulk modulus K increases with increasing density n for all three lattices and diverges upon close-packing.

As a function of the screening length at fixed density ($n = 0.9$), the bulk modulus and three cubic elastic constants are shown in Figs. 7.4 and 7.5, respectively. Though the elastic constants appear to peak when the inter-particle spacing is roughly on the order of three screening lengths, this is an artifact of our model: as we discussed in [11], when the interparticle spacing, a, is comparable to the Debye screening length, next to nearest neighbor interactions should be included. Since our Derjaguin-like approximation only accounts for nearest neighbors, it is not reliable for $\lambda \equiv \kappa a$ of order 1. Thus we expect that the peaks in the moduli mark the crossover at which the nearest neighbor approximation breaks down. We duplicate the analysis at densities $n = 0.7$ and 0.8, and we observe the same spurious maximum in the plots of the elastic constants but trust our results beyond the peak.

Next, we consider the elastic constants as functions of varying density at fixed screening length so that $\lambda = \kappa a = 4$ (Figs. 7.6, 7.7). Expectedly, the bulk modulus, as well as the three elastic constants K_{11}, K_{12}, and K_{44}, increase with increasing density. Moreover, they diverge upon approaching the close-packing limit. We expect this trend to be valid for all other

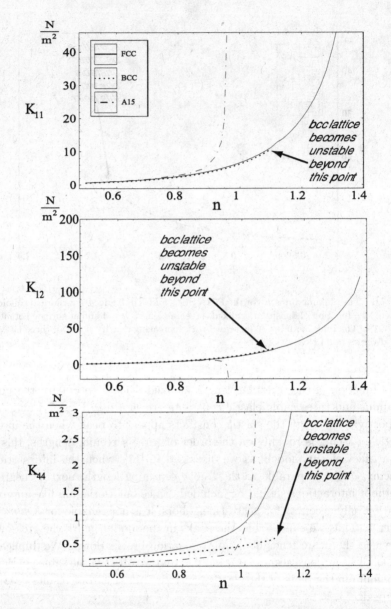

Fig. 7.7 The elastic constants K_{11}, K_{12}, and K_{44} for the FCC, BCC, and A15 lattices of a charged colloidal system as a function of density n, at fixed $\lambda = 4$ and at surface potential $\Phi_s = 0.4$. Similar to the bulk modulus K, the moduli of the elastic constants diverge upon close packing.

screening lengths as well. The constants K_{11} and K_{12}, which control the elongational shear mode, have similar magnitudes across different lattices, while the simple shear constant K_{44} is roughly one-tenth of the other two constants, dramatically smaller for the same systems. It follows that the shear moduli of the charged colloids are considerably smaller than the bulk moduli. If we consider the results for K_{44}, for instance, we see that the modulus is largest in the FCC lattice, and significantly smaller for the A15 lattice. This is similar to the case of hard spheres for which it is well known that the BCC lattice is unstable with respect to shear along the four-fold axis. It is conceivable that for the non-close packed lattices, there exist pockets of unoccupied space that become available upon shear which would allow an overall lowering of the elastic free energy. The effects of lattice distortion are less noticeable in the BCC and A15 lattices due to the greater availability of space for each lattice site. Physically, this property translates into the observed fact that the BCC and A15 lattices are much softer and much more amenable to shear deformations. The softening of the BCC shear moduli in comparison to the FCC ones has in fact been seen in MD studies of the FCC lattice of copper [12] and of the BCC lattice of vanadium [13].

In the case of the A15 lattice (Fig. 7.8), there is an interesting shear mode. This lattice has two distinct sites: pairs of columnar sites are located at bisectors of the faces of the unit cell, forming three sets of mutually perpendicular columns, and interstitial sites are at the vertices and at the center of the unit cell. Figure 7.8a shows a side view of the A15 lattice in the close-packing limit, which is determined by the density at which the columnar spheres touch each other — the interstitial sites are still far from their neighbors. There is a shear mode that can exploit this excess volume: consider the deformation shown in Fig. 7.8a along the {110} direction. The sheared lattice (dashed line) is to be contrasted with the cubic close-packing arrangement (solid line). The columnar sites now no longer touch one another, and the free energy is thus lowered in this configuration. However, as the shear increases, the columnar sites touch the interstitial sites and there can be no more distortion. Moreover, since the relative free volume of the columnar and interstitial sites depends on the density, this instability occurs only for a range of densities near the close-packing limit of the A15 lattice; the hard-sphere lattice becomes unstable with respect to shear along the face diagonal at densities greater than $n \approx 0.88$. This is qualitatively different from the shear instability of the BCC lattice, which is unstable at all densities.

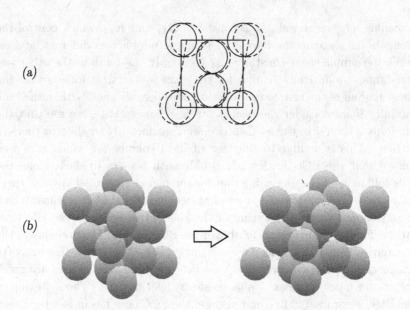

Fig. 7.8 (a) Schematic of the shear instability of the A15 lattice: the cubic arrangement of the spheres (solid line) is unstable to shear along the face diagonal (dashed line). (b) The cubic A15 lattice and the triclinic derivative lattice as the entropically preferred structure at high densities.

The shear instability shown in Fig. 7.8a only affects a fraction of all the sites in the unit cell and breaks the symmetry of the cubic lattice, rendering the three columnar sites inequivalent. Since there is no mechanism that would prefer shear along either of the three faces of the A15 unit cell, the actual instability experienced by the system should preserve the equivalence of the columnar sites. Thus it should include equal amounts of shear along all three faces, which corresponds to shear along the body diagonal. The resulting structure belongs to the triclinic system, since the edges of the sheared unit cell are no longer perpendicular to one another. However, the sheared lattice is a special triclinic lattice because the edges of the unit cell are of equal length and the angles between them are identical. It turns out that the angle between the edges which maximizes the packing fraction is $\arccos(1/4) = 75.5$ degrees (Fig. 7.8b). This configuration gives a packing fraction of 0.5700, to be compared with the A15 packing fraction of $\pi/6 = 0.5236$. Both this instability and the instability of the BCC lattice can be stabilized by the soft repulsion between the colloidal particles, encoded in the surface contribution to the elastic free energy. In both the

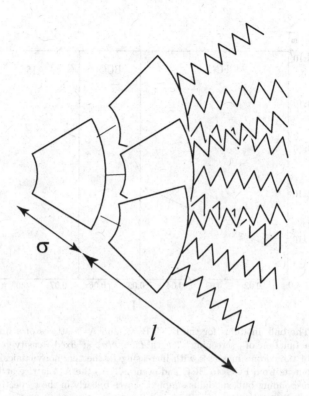

Fig. 7.9 The effective dimensionless "screening length" of the dendrimers, $L = \ell/\sigma$, corresponds to the ratio of the thickness of the soft outer part, consisting mainly of the alkyl corona, to the radius of the impenetrable hard core of the dendrimer molecule consisting of rigid aromatic rings.

fuzzy and charged colloidal systems considered in this paper, the parameters of the soft repulsive tail are such that the A15 lattice is stable at all densities considered.

7.4 Elasticity of Fuzzy Colloids

We now apply our foam model to the case of fuzzy colloids, of which the FCC and BCC lattices are the main candidates for the solid phase. For charged colloids, the screening length is the Debye screening length κ^{-1}; the analogous concept for the fuzzy colloid is the corona thickness (Fig. 7.9).

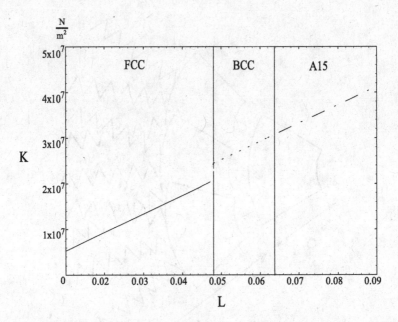

Fig. 7.10 The bulk modulus for the FCC, BCC, and A15 lattices of a fuzzy colloidal system as a function of "screening" length $L = \ell/\sigma$, at fixed density $n = 0.9$. As the length of the corona increases with increasing L, the thermodynamically favorable structure transits from FCC, to BCC and eventually to the A15 lattice at high enough L. The corresponding bulk modulus simply behaves linearly in the respective regions.

As before, we consider the effect of both varying the "screening" length and the density. At a fixed density of $n = 0.90$, we compare the total free energies within the high-density regime as done in Ref. [14] and find that the BCC lattice becomes more favorable than FCC at $\ell \approx 0.043\sigma$, whereas the BCC–A15 transition occurs at $\ell \approx 0.062\sigma$. In order to make comparison with the previous study, we will focus on dendrimers with $N_0 = 162$ dodecyl chains per 3rd and 4th generation micelle with a particle size of $\sigma = 4$ nm [15, 16].

The bulk modulus and the elastic constants are calculated for each lattice in their respective stability region at fixed density $n = 0.9$ and the results are shown in Figs. 7.10 and 7.11. Due to the functional form of the surface energy (Eq. (5.12)), the linear dependence on $L = \ell/\sigma$ is preserved in the bulk modulus and the three elastic constants. Thus they all vary directly with the screening length. We again observe the trend that K_{44} is smallest for the A15 lattice than for the other lattices, while the overall magnitude of K_{44} is less than those of K_{11} and K_{12}. Presumably,

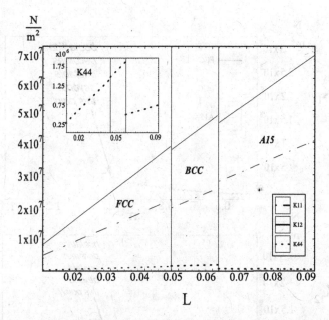

Fig. 7.11 The elastic constants K_{11}, K_{12}, and K_{44} for the FCC, BCC, and A15 lattices of a fuzzy colloidal system as a function of "screening" length $L = \ell/\sigma$, at fixed density $n = 0.9$. As the "screening" length L increases, non-close packed structures are favored. Similar to the bulk modulus, the three elastic constants behave linearly with respect to L.

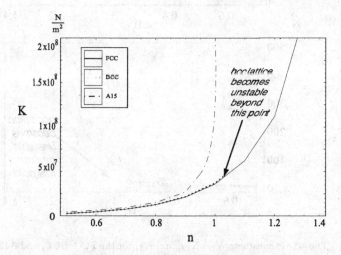

Fig. 7.12 The bulk modulus K for the FCC, BCC, and A15 lattices of a fuzzy colloidal system as a function of density n, at "screening" length $L = \ell/\sigma = 0.05$. As expected, the bulk modulus K increases with increasing density n.

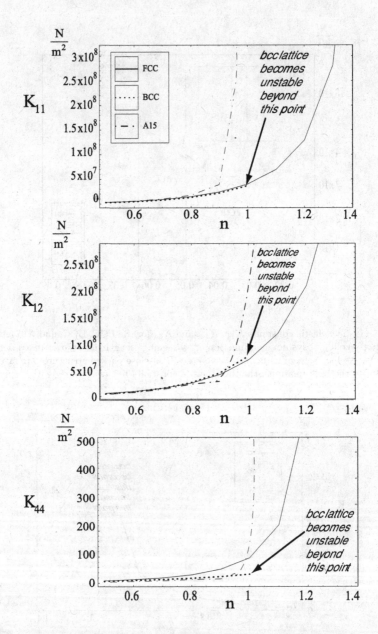

Fig. 7.13   The elastic constants K_{11}, K_{12}, and K_{44} for the FCC, BCC, and A15 lattices of a fuzzy colloidal system as a function of density n, at "screening" length $L = \ell/\sigma = 0.05$. They increase with increasing density n much like the bulk modulus.

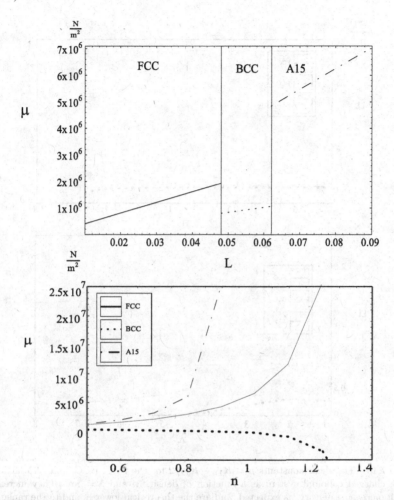

Fig. 7.14   The shear constants $\mu = K_{11} - K_{12}/2$ for the FCC, BCC, and A15 lattices of a fuzzy colloidal system as a function of density n and "screening" length $L = \ell/\sigma$. They result from the elongational shear mode and represent the lower bound of the average shear measurable in experiments.

the argument presented above for the charged colloids would apply here as well.

At fixed $L = 0.05$, the plots of the bulk modulus and the elastic constants as functions of density are shown in Figs. 7.12 and 7.13. Except for K_{12} in the A15 lattice, they all increase with density and diverge upon approaching close packing. The K_{44} values are also smaller than those of K_{11} and K_{12} across the lattices. The elongational shear constants for the

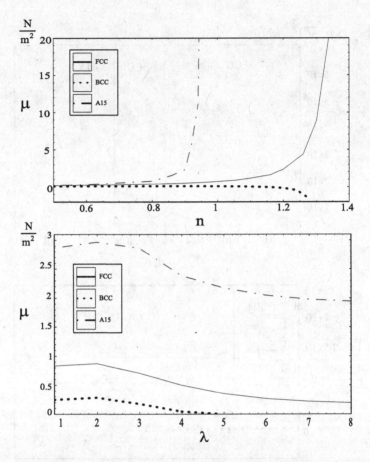

Fig. 7.15   The shear constants $\mu = K_{11} - K_{12}/2$ for the FCC, BCC, and A15 lattices of a charged colloidal system as a function of density n and $\lambda = \kappa a$. They increase with increasing density, as expected, and are the theoretical lower bound to the range of average shear.

non-close packed structures are consistently smaller than their FCC counterpart, in both systems of fuzzy and charged colloids. (Figs. 7.14 and 7.15). As with the charged colloids, the shear modulus of the BCC lattice becomes negative at large enough reduced density in the fuzzy colloids as well; only the physically relevant elastic constants have been included in our plots.

In summary, the elastic properties of the fuzzy colloids behave similarly to those of charged colloids, in spite of the overall scaling of the elastic constants. The six orders of magnitude difference between the two systems is simply due to the large difference in the sizes of fuzzy and charged colloidal particles.

7.5 Conclusion

To conclude, we have used the physical analogy with foams to study the phases of crystalline lattices as well as their elasticity properties. We find that while the bulk moduli of the lattices considered are quite similar at all densities away from their respective close packing limits, the relative differences of their shear moduli are much larger. The shear moduli of the non-close-packed lattices are smaller than those of the FCC lattice.

In many ways, the two systems we considered are similar, which would indicate that the details of the interparticle interaction are not essential for all macroscopic properties of colloids. We hope that the foam analogy used here provides an intuitive and mathematically tangible formalism that relates macroscopic properties of colloidal systems directly to their microscopic constituents. Recently, such materials are becoming increasingly important for a range of applications. We are hopeful that our model will serve as a useful guide for further study and synthesis and, possibly, the engineering of colloidal crystals of specific geometry and desired properties. Materials with increasingly more complex crystal structures are synthesized [17] and theoretically predicted [18] at an ever growing rate, and their understanding in terms of a heuristic model should prove useful.

References

[1] E. B. Sirota, *et al.*, *Phys. Rev. Lett.* **62**, 1524 (1989).

[2] J. F. Joanny, *J. Col. Int. Sci.* **71**, 622 (1979).

[3] S. Alexander, P. M. Chaikin, P. Grant, G. J. Morales, P. Pincus, and D. Hone, *J. Chem. Phys.* **80**, 5776 (1984).

[4] J. C. Crocker and D. G. Grier, *Phys. Rev. Lett.* **73**, 352, 1994.

[5] P. M. Chaikin and T. C. Lubensky, *Principles of Condensed Matter Physics*, Cambridge University Press, New York, 1980.

[6] K. Brakke, *Exp. Math.* **1**, 141 (1992).

[7] M. O. Robbins, K. Kremer and G. S. Grest, *J. Chem. Phys.* **88**, 3286 (1988).

[8] T. T. Nguyen, A. Yu. Grosberg, and B. I. Shklovskii, *Phys. Rev. Lett.* **85**, 1568 (2000).

[9] M. Boström, D. R. M. Williams, and B. W. Ninham, *Phys. Rev. Lett.* **87**, 168103 (2001).

[10] A. P. Gast, C. K. Hall, and W. B. Russel, *Faraday*.

[11] W. Kung, P. Ziherl and R. D. Kamien, *Phys. Rev. E* **65**, 050401(R) (2002).

[12] A. Kanigel, J. Adler and E. Polturak, *Int. J. Mod. Phys. C* **12**, 727 (2001).

[13] V. Sorkin, E. Polturak and J. Adler, *cond-mat/0304215* (2003).

[14] P. Ziherl and R.'D. Kamien, *J. Phys. Chem. B* **105**, 10147 (2001).

[15] V. S. K Balagurusamy, G. Ungar, V. Percec and G. Johansson, *J. Am. Chem. Soc.* **119**, 1539 (1997).

[16] P. Ziherl and R. D. Kamien, *Phys. Rev. Lett.* **85**, 3528 (2000).

[17] G. Ungar, Y. Liu, X. Zeng, V. Percec and W. D. Cho, *Science* **299**, 1208 (2003).

[18] C. N. Likos, N. Hoffmann, H. Löwen and A. A. Louis, *J. Phys. C* **14**, 7681 (2002).

PART 4

Geometry and Phase Transitions, in Topologically Constrained Polymers

Chapter 8

Topologically-Constrained Polymers in Theta Solution

8.1 Introduction

The statistics and properties of random walks are central to our understanding of a large class of phenomena in mathematics, biology, physics, and even economics. The wide applicability of random walks in modeling physical situations is well known is in part due to the Gaussian statistics of its correlations, rendering it attractive from an analytical viewpoint. In polymer physics, the model of random walks can be adapted to the study of polymers that are much longer their persistence length. Field-theoretic methods in the random-walk problem then translate into a theoretical basis for computation of scaling exponents relevant for polymer systems. One exponent of interest, μ, relates the radius of gyration R_G to the length of the chain L through $R_G \sim L^{\mu}$. In the case of long polymers in good solvents, the interactions of chain connectivity and of excluded volume are described by de Gennes' mapping [1] on the $O(N)$ symmetric ϕ^4-theory in the $N \to 0$ limit. Using renormalization-group analysis, one finds that the Wilson-Fisher point controls the scaling in $d = 4 - \epsilon$ dimensions, where the radius-of-gyration exponent ν and anomalous dimension γ ($Z \sim e^{\mu N} N^{\gamma-1}$) take the following values in $d = 3$: $\nu = 0.588$ and $\gamma = 1.157$. In the opposite limit of polymers dissolving in poor solvents, the monomer-monomer interaction becomes attractive and the polymers for compact globules with $R_G \sim L^{1/d}$. The transition between the two solvent regimes is known as Θ-transition at some temperature T_θ.

At the vicinity of the Θ-transition, the ϕ^4 term vanishes and makes the transition a critical point. The next-order potential term of ϕ^6 becomes important in the stability of the system in this scaling regime. As a result, de Gennes [2] proposed a mapping on the $O(N)$-symmetric ϕ^6-model, ascribing a second-order behavior to the Θ-transition. Logarithmic corrections [3, 4] to mean-field results have been calculated, but it has been proven difficult to test these predictions experimentally. Computer simulations [5–7] have also been performed but agreement to theory has only been tentative.

So far we have considered only free polymers in the sense that they have no topological constraint other than connectivity. Many polymers in nature, such as actin or DNA, have additional constraints such as twist rigidity. Refinements can be made to de Gennes' mapping to incorporate these additional variables. For the class of closed-loop polymers, the $N \to 0$ limit of a $U(N)$ Chern-Simons theory with the gauge field coupled to the common $U(1)$ is considered. In this mapping, the abelian Chern-Simons gauge field is used to enforce the topological constraint in the form of White and Fuller's [8, 9] famous relation $Lk = Tw + Wr$ [13, 10–12]. It relates the topological linking Lk to the total amount of twist Tw and writhe Wr found in, for example, DNA backbones. The consideration of such constraint becomes important as recent progress in experimental techniques has enabled the direct, real-space observation of polymer confirmations, and coupling was discovered [14] between excess in linking number and backbone configuration in studies of DNA force-extension curves. The topological constraint imposed by Lk is inherently three-dimensional as is the Chern-Simons term which surrogates for it. As a result, the perturbative study of a polymer in good solvent cannot be performed via a conventional $\epsilon = 4 - d$ expansion. In [15] it was shown that the exponents at the $d = 3$ Wilson-Fisher fixed point were unchanged at one-loop order. Because there is no controlled expansion to go to higher order, we have focussed here on polymers at the Θ-point with a topological constraint so that a systematic expansion can be performed directly in $d = 3 - \epsilon$ dimensions.

8.2 $O(N)$-Symmetric ϕ^6-Theory

The simplest model exhibiting tricritical phenomenon is the scalar ϕ^6-theory. In general, the existence of multicritical points is due to the interaction between two or more order parameters of the system. Specifically, the crossover behavior between different phases is determined by the competition between the divergences of different correlation length scales associated with these order parameters [16].

In nature, examples of tricritical phenomenon abound. Among the notable physical realizations are the He^3–He^4 system of binary mixtures [17] or the absorption of helium on krypton-plated graphite [18]. In particle physics, the ϕ^6-theory is used in the study of polarons, the soliton-like field configurations found in low-dimensional systems [19]. Recently, the introduction of a $SU(2)$-gauge to the theory in three-dimensional Euclidean space was used to study a possible tricritical point in the Higgs models at high temperature [20]; in this case, the electroweak phase transition is of interest. Going to two dimensions, an alternative description of tricritical phenomena avails in terms of conformal field theories: every unitary minimal model can be associated with the critical limit of a lattice model [21, 22]. For example, the lattice tricritical Ising model corresponds to the next minimal model $\mathcal{M}_4$ with central charge $c = 7/10$ or the simplest $N = 1$ superconformal minimal model $\mathcal{SM}_3$ [23]. In general, for the $\mathcal{M}_m$ model of primary fields with central charge $c = 1 - 6/m(m+1), m = 3, 4, \ldots$, there associates the action of an equivalent Landau-Ginzburg description in terms of scalar fields:

$$S_{LG} = \int \mathrm{d}^2 z \frac{1}{2} (\partial \phi)^2 + \phi^{2m-2}. \tag{8.1}$$

Likewise, the $N = 1$ superconformal unitary minimal series $\mathcal{SM}_n$ with central charge $c = 3/2 - 12/n(n+2), n = 3, 4, \ldots$, has the following equivalent description in terms of superfields:

$$S_{LG}^{N=1} = \int \mathrm{d}^2 z \, \mathrm{d}^2 \theta \, \frac{1}{2} (D\Phi)^2 + \Phi^n. \tag{8.2}$$

Therefore in two dimensions, critical phenomena near the tricritical point can be studied as either a bosonic theory with a ϕ^6 potential or as a $N = 1$ supersymmetric theory with a Φ^3 potential [24]. In our context of polymers, we mainly focus on the bosonic description of the phenomenon, which works in arbitrary physical dimensions.

For polymer systems, while chemists and biologists deal with the local properties of the systems such as the molecular structure of individual monomers and their chemical interactions, physicists focus on the global properties of the polymers without regard to its specific composition. To this end, we note that a typical monomer prefers to be surrounded by other monomers, as supposed to solvent particles, when dissolved in solution. This preference creates an attractive potential between the monomers, akin to a long-range van der Waals force. Physically, the two monomers cannot overlap each other in space, and this physical constraint gives rise to a hard-core repulsion. At different temperatures, the balance between

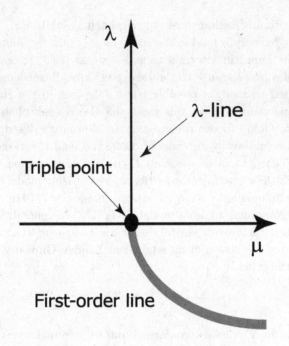

Fig. 8.1 Phase Diagram for a Tricritical System. The line $\mu = 0$, $\lambda > 0$ is a second-order lambda line, shown as a single line in the figure. The line $\mu = \frac{1}{2}|\lambda|^2/\nu$ is a line of first-order transition, shown as a thickened grey line in the figure. The triple point occurs at $\mu = 0$, $\lambda = 0$.

these two interactions determines the behavior of the whole system. At high temperatures, one can readily see the hard-core repulsion as the dominant interaction, whereas the attractive potential dominates at low temperatures. The condition of good solvent occurs in the former scenario, and that of poor solvent characterizes the latter. Expectedly, there must exist a special temperature at which the attractive potential exactly cancels the effect of the hard-core repulsion. Polymers in this condition behave ideally as there is no penalty for monomer-monomer contacts. This condition defines the Θ-point of the system. The corresponding Θ-transition is second-order in nature (Fig. 8.1).

To establish the connection between random walks and critical behavior of polymers, we consider the discrete Edwards model, in which the configuration of a polymer chain is described as a continuous curve $r^a(s)$ in d-dimensional space, where s is the arc length of the polymer and a denotes the a-th component of r. To adequately describe the physical behavior of the polymer chain, the interaction energy is taken to be the following form:

$$E(r^a) = \frac{1}{4} \int_0^s ds \left(\frac{dr^a(s)}{ds} \right) + \frac{u_0}{4!} \int_0^s ds_1\, ds_2\, \delta^d \left(r^a(s_1) - r^a(s_2) \right)$$

$$+ \frac{w_0}{6!} \int_0^s ds_1 ds_2 ds_3\, \delta^d \left(r^a(s_1) - r^a(s_2) \right) \delta^d \left(r^a(s_2) - r^a(r_3) \right).$$

$$(8.3)$$

The first term describes the connectivity of the chain, whereas the second and third terms are local two-body and three-body interaction of strengths u_0 and w_0, respectively. In describing critical phenomena the field-theoretic approach is often desirable, since the correlation lengths of relevant degrees of freedom become very long, and the discrete structure is no longer seen in this limit. Mapping Eq. (8.3) to a local field theory via Laplace transform [1, 2], we arrive at the Landau-Ginzburg description of the system in terms of a scalar field ϕ. At the Θ-point, the exponent ν is the correlation length exponent at the tricritical point $\mu = u = 0$ of the following free energy density for the N-component, complex scalar $\vec{\phi}$:

$$f = |\partial_\mu \phi_i|^2 + \mu |\vec{\phi}|^2 + u \left(|\vec{\phi}|^2 \right)^2 + w \left(|\vec{\phi}|^2 \right)^3, \qquad (8.4)$$

in the $N \to 0$ limit. The $N \to 0$ limit can be derived from the relation between the self-avoiding random walks and the high-temperature Ising model for spins on a lattice. Starting with the partition function of the N-component Ising model,

$$z = \sum_{\{\sigma_i\}} \exp \left[\beta \sum_{\{i,j\}} \sigma_i^\alpha \sigma_j^\beta \right] \qquad (8.5)$$

where $\sigma_i^\alpha \sigma_i^\alpha = N$ and $\langle \sigma_i^\alpha \sigma_i^\beta \rangle = \delta^{\alpha\beta}$. On the square lattice, the *Feynman diagrams* in the high-temperature expansion of the above Hamiltonian corresponds exactly to the geometry of random walks (Fig. 8.2). In a random walk, the expectation value of the distance R traversed by the walker after N steps has the defining value of $\langle R_N \rangle = 0$. The expectation value of the second moment, however, is non-vanishing and can be computed as follows:

$$\langle R_N^2 \rangle = \langle R_{N-1}^2 \rangle + l^2 \qquad (8.6)$$

$$\langle R_N^2 \rangle = N l^2 \qquad (8.7)$$

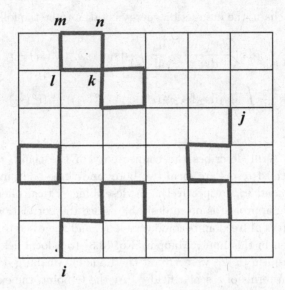

Fig. 8.2 Some possible paths for the high-temperature expansion of the Ising model for spins on a square lattice. While the path $(i \to j)$ is self-avoiding, the path $(k \to l \to m \to n)$ is self-intersecting and vanishes in the $N \to 0$ limit.

since $\langle R_0^2 \rangle = 0$. As shown in Fig. 8.2, when a path intersects itself, it necessarily forms a closed loop $(k \to l \to m \to n)$. When summed over indices, such closed loops also necessarily involves factors of N. As a result, the limit $N \to 0$ leads to the elimination of all such self-intersecting paths and the sole existence of self-avoiding paths $(i \to j)$. A more rigorous derivation in terms of correlation functions of the fields is also possible [25].

The thermodynamics of the theory can be determined from Eq. (8.4) as follows. If the four-point coupling u is positive, the sixth-order term can be neglected in the vicinity of the second-order transition. If u is negative, the sixth-order term is needed to maintain stability of the system. From dimensional analysis, the ϕ^6-theory is renormalizable in three dimensions. Mean-field behavior is predicted at the tricritical point, and logarithmic corrections modify scaling in the critical region [3, 26, 27]. Recently these corrections have been calculated to six-loop order [28, 29].

Schematically, one can employ the BPHZ renormalization scheme in the computation of divergent Feynman diagrams. First, we introduce the following renormalized free energy:

$$F = \int \mathrm{d}^3 x \, Z_\phi |\partial_\mu \phi_i|^2 + \tilde{\mu} Z_\mu |\vec{\phi}|^2 + Z_w \tilde{w} \left(|\vec{\phi}|^2 \right)^3 , \qquad (8.8)$$

where $\tilde{\mu} = \mu M^2$ and $\tilde{w} = wM^{6-2d}$, and M is the momentum scale at which we are renormalizing. Let $\Gamma^{(N,N_{\phi^2})}$ be the vertex function with N external legs and N_{ϕ^2} insertions of ϕ^2. Furthermore, let $\bar{R}$ and K be the Bogoliubov's $\bar{R}$-operation of recursively subtracting the divergences of all subdiagrams and the operation of selecting the pole of the Laurent series expansion in ϵ, respectively. We have the following expressions for the various Z factors:

$$Z_\phi = 1 - \frac{\partial}{\partial p^2} K \bar{R} \Gamma^{(2,0)}(p,\mu,w) \tag{8.9}$$

$$Z_\mu = 1 - \frac{\partial}{\partial \mu} K \bar{R} \Gamma^{(2,0)}(p,\mu,w) \tag{8.10}$$

$$+ Z_w = 1 - \frac{1}{w} K \bar{R} \Gamma^{(6,0)}(p,\mu,w) \;. \tag{8.11}$$

A convenient way to regularize the divergent Feynman amplitudes is the method of dimensional regularization. An ϵ-expansion is carried out about the critical dimension $d = 3$. The engineering dimension of any correlation function Γ depends on its field content so that $d(\Gamma) = 3 - \frac{1}{2}N_\phi$. The Callan-Symanzik equation for the $O(N)$-symmetric ϕ^6-model is given by:

$$\left[M\frac{\partial}{\partial M} + \beta_\mu \frac{\partial}{\partial \mu} + \beta_w \frac{\partial}{\partial w} - \frac{1}{2}N_\phi \eta_\phi \right] \Gamma_R^{(N_\phi)} = 0, \tag{8.12}$$

where M is the renormalization scale, and the coupling constants and renormalization functions are given by the following:

$$\eta_\phi = M\frac{\partial}{\partial M} \ln Z_\phi \,, \tag{8.13}$$

$$\mu_0 = \mu M^2 \frac{Z_\mu}{Z_\phi} \,, \tag{8.14}$$

$$w_0 = wM^{6-2d} \frac{Z_w}{Z_\phi^3} \,. \tag{8.15}$$

The first-order corrections to the mean-field results is shown in Fig. 8.3. It is a straightforward albeit laborious exercise in computing the Feynman diagrams and extracting the renormalization constants. Taking into account various symmetry factors arising from different contractions of ϕ with $\phi*$ and summing over the $O(N)$ indices, the evaluation of Fig. 8.3a, fig. 8.3b

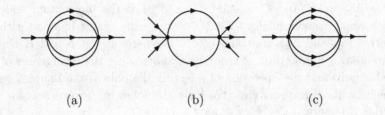

(a)   (b)   (c)

Fig. 8.3   First-order corrections to Z_ϕ (a), Z_w (b), and $Z_\mu(p = 0)$ (c) in $O(N)$-symmetric ϕ^6-theory. Note that figures (a) and (b) are already four loops in contribution.

and Fig. 8.3c gives the following expressions of Z_ϕ, Z_μ, and Z_w [28, 29], respectively, in the $N \to 0$ limit:

$$Z_\phi = 1 - \frac{3w^2}{8\epsilon\pi^4}, \tag{8.16}$$

$$Z_\mu = 1 + \frac{45w^2}{8\epsilon\pi^4}, \tag{8.17}$$

$$Z_w = 1 + \frac{33w^2}{4\pi^2\epsilon}. \tag{8.18}$$

The β-function of the six-point coupling w can be readily determined:

$$\beta_w = -w\left(2\epsilon - \frac{33w}{2\pi^2}\right), \tag{8.19}$$

which has a non-trivial infrared-stable fixed point at $w^* = 4\pi^2\epsilon/33$. In the limit of $\epsilon = 3-d \to 0$, this Wilson-Fisher fixed-point converges to the trivial fixed point at $w^* = 0$. Therefore, as expected, there are no corrections to the mean-field exponents in three dimensions. In particular, we have the following simple scaling law for the radius of gyration:

$$R_G = L^{1/2}. \tag{8.20}$$

8.3   Chern-Simons Theory and Writhe

In the previous section, we used the $O(N)$-symmetric ϕ^6-theory to compute the radius-of-gyration exponent ν and established that fluctuations near the Θ-point does not modify this scaling exponent of free polymers from its mean-field value in this critical regime. To incorporate the additional topological constraint of linking number to the polymers (Fig. 8.4), we

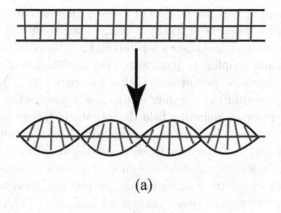

(a)

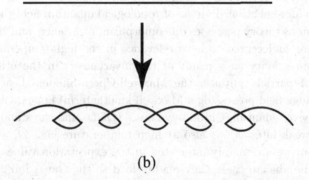

(b)

Fig. 8.4 Twist can be understood as the rotation of cross section about the tangent of the two backbones; the overall length remains constant in this mode of topological constraint (a); Writhe is simply the integrated torsion of a line; unlike twist, the projected length of the object changes in this mode (b).

modify the ϕ^6-theory by introducing an abelian Chern-Simons gauge field that couples covariantly to the common $U(1)$.

Logically, the Chern-Simons theory makes for a sensible model as its action is independent of the choice of metric: it encodes only topological information on the manifold upon which the theory is defined. In our case of polymers, the topological constraint of Lk in three-dimensional Euclidean space comes out naturally when one constructs correlation functions of objects fundamental to the theory; namely, *Wilson loops*.

Much like the ϕ^6-theory, the pure Chern-Simons theory has seen many diverse applications. The low-energy physics of the fractional quantum Hall effect is described by effective degrees of freedom known as anyons possessing the distinctive property of statistics transmutation [30–33]. These are quasi-particles whose spin-statistics depends on the Chern-Simons coupling. This Chern-Simons coupling is quantized in the non-abelian case and arbitrary in the abelian case permitting fractional statistics [34, 35].

The quantization of the coupling constant is a noteworthy feature indicative of a topological quantum field theory, of which the Chern-Simons theory is an example. The original motivation of its study lies in Witten's discovery [36] that the correlation functions of Wilson loops in the theory are topological invariants of the underlying three-dimensional manifolds. The field-theoretic formulation of the problem provides not only an intrinsically three-dimensional conceptual framework (which was previously lacking) but also an important formal computational tool in the classification theory of knots in arbitrary three-spaces.

By the aforementioned virtue of topological quantum field theories, the Chern-Simons theory possesses diffeomorphism invariance, and this class of theories are conjectured to have relevance in the high-temperature phase of quantum gravity as a result of the invariance. In the other cosmic extreme of particle physics, the Maxwell-Chern-Simons theory (Chern-Simons gauge field possessing a Maxwell kinetic term) appears an an effective theory of Quantum Chromodynamics (QCD) and the standard model of electroweak interactions also at high temperature [38, 39, 37]. In our application, we are mainly interested in the expectation values of Wilson loops in the abelian case: they readily lead to the Gauss linking number of two knots upon integration and provide the correct prescription of our topological twist constraint for the closed polymer loops.

In general, formulation of a quantum field theory requires a choice of Lagrangian and a set of suitable observables, with the necessity that symmetries of the Lagrangian must also be respected by the observables. The Chern-Simons theory is unique in that, in addition to the usual gauge invariance found in Yang-Mills theories, it also possesses general covariance. Yet unlike General Relativity, general covariance is achieved not by functionally integrating over all metrics but by the selection of a metric-independent gauge-invariant Lagrangian. With this in mind, the usual gauge invariant local operators would not be appropriate observables as they spoil general covariance. Though all is not lost, as one class of objects known as the Wilson lines, most familiar in the context of QCD, readily extend themselves as the appropriate choice of observables in our metric-independent context. A related concept known as the Wilson loops, which is more suitable to our goal of studying polymers, is simply the trace of the holonomy of the gauge field around a loop in space:

$$W(C) = TrP \exp \oint_C A_\mu \, dx^\mu, \tag{8.21}$$

Physically, the Wilson loop is a modern reincarnation of Faraday's notion of field lines in electromagnetism. Given a quantized gauge field A_μ, the Wilson loop is just an operator on the Hilbert space of states; applying this operator to the vacuum leads directly to a quantum state in which the Yang-Mills analog of the electric field flows around the loop. The Wilson loops ensure unbroken general covariance, which is important in extracting global topological properties of the system, as the lack of a metric prohibits signal propagation and thus there is no local physics.

To see the connection of the Chern-Simons theory to the topological constraint of writhe, we begin with the abelian Chern-Simons Lagrangian [40]:

$$\mathcal{L}_{CS} = \frac{k}{8\pi} \int_M \epsilon^{ijk} A_i \partial_j A_k. \tag{8.22}$$

Choosing M to be circles C_a and some integers n_a that correspond to representations of the gauge group $U(1)$, and assuming that there is no intersection between C_a and C_b for $a \neq b$, the product of Wilson loops simple becomes:

$$W = \prod_{a=1}^{s} \exp \left(i n_a \int_{C_a} A \right). \tag{8.23}$$

The correlation in Eq. (8.23) can be computed accordingly with respect to the Gaussian measure of the $\exp(i\mathcal{L})$ phase factor. The end result is the Gauss linking number:

$$\langle W \rangle = \exp \left(\frac{i}{2k} \sum_{a,b} n_a n_b \int_{C_a} dx^i \int_{C_b} dy^j \, \epsilon_{ijk} \frac{(x-y)^k}{|x-y|^3} \right). \tag{8.24}$$

The integral within the exponential, modulus numerical factors, is the definition of the writhe $Wr[R]$ of curve R:

$$Wr[R] = \frac{1}{4\pi} \int ds \int ds' \left(\frac{dR(s)}{ds} \times \frac{dR(s')}{ds'} \right) \cdot \frac{R(s) - R(s')}{|R(s) - R(s')|^3}. \tag{8.25}$$

It can be seen from Eq. (8.25) that writhe is a non-local concept.

Now that we see how the geometrical concept of writhe can arise out of the correlation functions of Wilson loops in the abelian Chern-Simons theory, we can relate it to the more general concept of the linking number Lk, via the White and Fuller's relation:

$$Lk = Tw + Wr. \qquad (8.26)$$

Experimentally, the most direct way to introducing topological constraints to polymers is to create a chemical potential for the linking number, rather than that for the writhe. For example, DNA plasmids in the presence of topoisomerase and ATP would tend to have some non-zero excess linking number due to the bias of the added enzyme. Fortunately, it can be shown mathematically that a chemical potential for the linking number is equivalent to a chemical potential for writhe [15]. To do so, we first assume that the free energy of the polymer is determined by a bending stiffness κ and twist rigidity C in the long-wavelength limit, as follows [13]:

$$F[R, \Omega(s)] = \frac{1}{2} \int ds\, \kappa \left(\frac{d^2 R}{ds^2} \right)^2 + C\Omega^2, \qquad (8.27)$$

where $R(s)$ describes the conformation of the chain backbone, $\Omega(s)$ is the twist degree of freedom, and s is the arc length. Using Eq. (8.26), the partition function for this model with a chemical potential g^2 for the linking number then becomes:

$$Z = \int [dR][d\Omega] \, \exp \left\{ -F[R, \Omega] - g^2 \left(Wr[R] + \frac{1}{2\pi} \int \Omega \, ds \right) \right\}. \qquad (8.28)$$

Thus, one can readily see that the twist degree of freedom can be integrated out to yield a new effective partition function with only a chemical potential for writhe:

$$Z_{eff} = \int [dR] \, \exp \left\{ -\frac{\kappa}{2} \int ds \left(\frac{d^2 R}{ds^2} \right)^2 - g^2 \, Wr[R] \right\}. \qquad (8.29)$$

In short, we show that the experimentally relevant chemical potential for linking number is equivalent mathematically to a chemical potential for writhe. And the latter can be obtained as correlation functions of the topological quantum field theory of Chern-Simons. To study the statistics of closed polymers obeying the White and Fuller relation in Eq. (8.26) near the Θ-point, we introduce an abelian Chern-Simons gauge field to the ϕ^6-theory and analyze the resulting theory using renormalization group techniques.

8.4 One-Loop Scaling of Closed Polymers

Since we are interested in the critical behavior of our theory near the Θ-point, we perform our analysis at the tricritical point where the couplings μ and u vanish. The energy is the sum $F = F_{\text{CS}} + F_{\text{b}} + F_{\text{gf}}$ where

$$F_{\text{CS}} = \frac{1}{2} \int d^3x\, \epsilon_{\mu\nu\rho} A_\mu \partial_\nu A_\rho \tag{8.30}$$

$$F_{\text{b}} = \int d^3x\, |\partial_\mu \phi_i - ig_0 A_\mu \phi_i|^2 + V(\phi_i) \tag{8.31}$$

$$F_{\text{gf}} = \frac{1}{2\Delta} \int d^3x (\partial_\mu A_\mu)^2 \tag{8.32}$$

$$V(\phi_i) = \mu_0 |\vec{\phi}|^2 + w_0 \left(|\vec{\phi}|^2 \right)^3 \tag{8.33}$$

where F_{CS} is the abelian Chern-Simons density, F_{b} is the scalar ϕ^6 term, and F_{gf} is the gauge-fixing term. From this point on we will only be interested in the Euclidean setting for our theory, so differentiation between vectors and forms and their respective index placements will no longer be relevant. We impose the Landau gauge ($\Delta \to 0$) in F_{gf} for all our subsequent calculations. Standard dimensional regularization is effective and leads to $\frac{1}{\epsilon}$ poles in $d = 3 - \epsilon$ dimensions. As is usual, we first perform the necessary tensor algebra in physical dimensions before analytically continuing the dimensions of the resulting scalar integrand. The correlation functions satisfy the Slavnov-Taylor identities [41] and the Ward identities [42]; gauge invariance is therefore preserved.

We first introduce the renormalized free energy as follows:

$$F = \int d^3x \left\{ Z_\phi |\partial_\mu \phi_i|^2 + \tilde{\mu} Z_\mu |\vec{\phi}|^2 - i\tilde{g} Z'_g \left[(A_\mu \phi_i)^* \partial_\mu \phi_i + (\partial_\mu \phi_i^*) A_\mu \phi_i \right] \right.$$

$$+ Z''_g \, \tilde{g}^2 \, |A_\mu \phi|^2 + \frac{1}{2} Z_A \, \epsilon^{\mu\nu\rho} A_\mu \partial_\nu A_\rho + \frac{1}{2\Delta} (\partial_\mu A^\mu)^2$$

$$\left. + Z_w \tilde{w} \left(|\vec{\phi}|^2 \right)^3 \right\}, \tag{8.34}$$

where $\tilde{g} = gM^{(3-d)/2}$, $\tilde{\mu} = \mu M^2$, $\tilde{w} = wM^{6-2d}$, and M is the momentum scale at which we are renormalizing. Denoting $\Gamma_R^{(N_\phi, N_A)}$ as the renormalized proper vertex of N_ϕ scalar fields and N_A gauge fields, the Callan-Symanzik equation is:

$$\left[M\frac{\partial}{\partial M} + \beta_\mu\frac{\partial}{\partial \mu} + \beta_w\frac{\partial}{\partial w} + \beta_g\frac{\partial}{\partial g} - \frac{1}{2}N_\phi\eta_\phi - \frac{1}{2}N_A\eta_A\right]\Gamma_R^{(N_\phi,N_A)} = 0,$$

(8.35)

where M is the renormalization scale. The engineering dimension of any correlation function Γ depends on its field content so that $d(\Gamma) = 3 - \frac{1}{2}N_\phi - N_A$. We have

$$\eta_\phi = M\frac{\partial}{\partial M}\ln Z_\phi \tag{8.36}$$

$$\eta_A = M\frac{\partial}{\partial M}\ln Z_A \tag{8.37}$$

$$\mu_0 = \mu M^2\frac{Z_\mu}{Z_\phi} \tag{8.38}$$

$$g_0 = gM^{(3-d)/2}\frac{Z_g'}{Z_\phi\sqrt{Z_A}} \tag{8.39}$$

$$w_0 = wM^{6-2d}\frac{Z_w}{Z_\phi^3}. \tag{8.40}$$

Figure 8.5 shows the diagrams arising from the Chern-Simons gauge field coupling to matter. Figures 8.5(a) and 8.5(b) are matter corrections to the gluon self-energy. Figures 8.5(c) and 8.5(d) are the gauge-field induced correction to matter self-energy. Figures 8.5(e) and 8.5(f) are the one-loop corrections to the gauge-field and matter coupling. We will only consider the infinite part in the evaluation of these diagrams, as it is these infinite parts that are independent of renormalization scheme and contribute to the β-function.

For the one-loop matter correction to the gauge-field self-energy, we obtain the following integrals:

$$\text{Fig. 8.5(a)} = (-g)^2\int\frac{d^dk}{(2\pi)^d}\frac{(2k+p)_\mu(2k+p)_\nu}{k^2(k+p)^2}, \tag{8.41}$$

$$\text{Fig. 8.5(b)} = -g^2\times N\int\frac{d^dk}{(2\pi)^d}\frac{g_{\mu\nu}\epsilon_{\mu A\nu}k_A}{k^2}. \tag{8.42}$$

The factor of N arises from summing up the indices for the closed scalar loops. In the $N \to 0$ limit found in de Gennes' mapping of random walks,

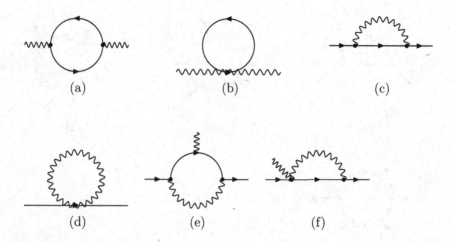

Fig. 8.5 One-loop diagrams arising from the gauge field coupling to matter: correction to the gauge field self-energy (a), (b); correction to matter self-energy (c), (d); correction to the three-point vertex (e), (f).

both of these diagrams vanish, and the gauge field receives no renormalization at one-loop level.

Generally, it has been established that the Chern-Simons term receives only finite renormalization in perturbation when it is coupled to massless matter beginning at two loops [43], while in the case of coupling to massive matter finite renormalization of the gauge field begins at one loop [44, 45].

In our specialized case of $N \to 0$ limit, a graph with only external gauge-field legs can receive no corrections. Since an external gauge-field leg must necessarily connect to two internal ϕ legs (and possibly one internal gauge-field leg) and since the $U(N)$ index of ϕ is not carried by any of the other external legs, there must be a sum over that index. Since that sum is proportional to N, this graph vanishes as $N \to 0$. More complex internal topologies only *add* factors of N to the graph and do not change this result, thus $Z_A = 1$. Note that Eq. (8.39) also vanishes by the antisymmetry of the gauge propagator.

For the one-loop gauge correction to the self energy of ϕ-field, we consider Figs. 8.5(c) and 8.5(d). From the first diagram we get the following expression;

$$\text{Fig. 8.5(c)} = (-g)^2 \delta^{ij} \int \frac{d^d k}{(2\pi)^d} \frac{\epsilon_{\mu\nu\lambda}(2p-k)_\mu k_\nu (2p-k)_\lambda}{k^2 (p-k)^2}, \qquad (8.43)$$

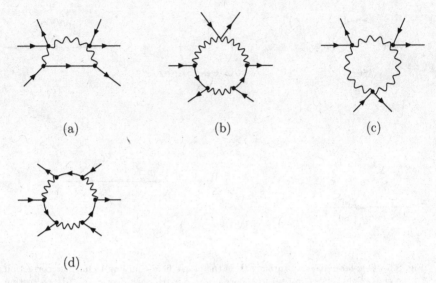

(a)   (b)   (c)

(d)

Fig. 8.6   One-loop contributions to $\Gamma^{(6,0)}$.

which again vanishes identically due to the antisymmetry of $\epsilon_{\mu\nu\lambda}$. Fig. 8.5(d) contains a four-point $\phi\phi AA$-vertex and vanishes identically in the same manner.

The remaining diagrams correct the three-point $\phi\phi A$-vertex to one-loop order. Their corresponding expressions are as follows:

$$\text{Fig. 8.5(e)} = (-g)^3 \delta^{ij} \int \frac{d^d q}{(2\pi)^d} \frac{\epsilon_{\mu\nu\beta}(2p-q)_\alpha(2p-2q+k)_\delta}{(p-q)^2 q^2}$$

$$\times \frac{q_\gamma(2p+2k-q)_\beta}{(p+k-q)^2} \tag{8.44}$$

$$\text{Fig. 8.5(f)} = (-g)^3 \delta^{ij} \int \frac{d^d q}{(2\pi)^d} \frac{g_{\mu\alpha}\epsilon_{\alpha\nu\beta}q_\nu(2p+2k-q)_\beta}{q^2(p+k-q)^2}. \tag{8.45}$$

It turns out that Eq. (8.45) is a finite expression, which does not contribute to either the β-function or the anomalous dimension of ϕ, while Eq. (8.45) vanishes identically.

There are four more diagrams giving gauge corrections to the six-point w-coupling (Fig. 8.6). Their corresponding expressions are the following:

$$\text{Fig. 8.6.x} \propto \delta^{ab}\delta^{cd}\delta^{ef} \int \frac{d^d q}{(2\pi)^d} \frac{N_x}{D_x}; \quad x = a, b, c, d \tag{8.46}$$

where

$$N_a = (-g)^2(-g^2)^2, g_{\gamma\delta}\, g_{\alpha\beta}\, \epsilon_{\nu A\gamma}\, \epsilon_{\delta B\alpha}\, \epsilon_{\beta C\mu}(2p_1 + k)_\mu(p_1 + k - p_2)_\nu$$
$$\cdot (p_1 + p_2 + k)_A(p_1 + p_2 + p_3 + k)_B k_C \tag{8.47}$$

$$D_a = (p_1 + k)^2(p_1 + p_2 + k)^2(p_1 + p_2 + p_3 + k)^2 k^2 \tag{8.48}$$

$$N_b = (-g)^4(-g^2)\, g_{\gamma\beta}\, \epsilon_{\alpha A\beta}\, \epsilon_{\gamma B\delta}\, \epsilon_{\nu C\mu}\,(2p_1 + k)_\mu(p_1 + k - p_2)_\alpha$$
$$\cdot (p_1 + p_2 + k)_A(p_1 + p_2 + p_3 + k)_B k_C(p_1 + p_2 + p_3 + 2p_4 + k)_\delta$$
$$\cdot (2p_1 + 2p_2 + 2p_3 + 2p_4 + k)_\nu \tag{8.49}$$

$$D_b = (p_1 + k)^2(p_1 + p_2 + k)^2(p_1 + p_2 + p_3 + k)^2$$
$$\cdot (p_1 + p_2 + p_3 + p_4)^2 k^2 \tag{8.50}$$

$$N_c = (-g^2)^3\, g_{\mu\alpha}\, g_{\beta\gamma}\, g_{\nu\delta}\, \epsilon_{\alpha A\beta}\, \epsilon_{\delta B\delta}\, \epsilon_{\nu C\mu}(p_1 + k)_A(p_1 + k + p_2)_B k_C \tag{8.51}$$

$$D_c = (p_1 + k)^2(p_1 + p_2 + k)^2 k^2 \tag{8.52}$$

$$N_d = (-g)^6\, \epsilon_{\alpha A\beta}\, \epsilon_{\gamma B\delta}\, \epsilon_{\mu C\nu}(p_1 - k)_\alpha(p_1 + k)_A(p_1 + k + 2p_2)_\beta$$
$$\cdot (p_1 + p_2 - p_3 + k)_\gamma(p_1 + p_2 + p_3 + k)_B(p_1 + p_2 + p_3 + 2p_4 + k)_\delta$$
$$\cdot (p_1 + p_2 + p_3 + p_4 - p_5 + k)_\mu(p_1 + p_2 + p_3 + p_4 + p_5 + k)_C$$
$$\cdot (p_1 + p_2 + p_3 + p_4 - p_5 - k)_\nu \tag{8.53}$$

$$D_d = (p_1 + k)^2(p_1 + p_2 + k)^2(p_1 + p_2 + p_3 + k)^2(p_1 + p_2 + p_3 + p_4 + k)^2$$
$$\cdot (p_1 + p_2 + p_3 + p_4 + p_5 + k)^2 k^2 \,. \tag{8.54}$$

As before, we are only interested in terms that can potentially contribute to the logarithmic divergences. By power counting, the leading terms of all four diagrams have the structure of logarithmic divergence followed by finite contributions. Fortunately, the antisymmetry of the gauge propagator once again renders these leading divergent terms identically zero. Therefore, $\Gamma^{(6,0)}$ and the coupling w is not renormalized by the Chern-Simons gauge field.

8.5   Two-Loop Results

In the previous section, we showed that there was no divergence in any of the one-loop Feynman diagrams. As a result, the mass term ϕ^2 and coupling w do not receive renormalization at one loop, and the radius-of-gyration exponent ν still takes on its mean-field value of $1/2$. This finding is not surprising, as one examines the symmetry of our symmetric $U(N)$-Chern-Simons Lagrangian.

Perturbatively, the underlying symmetry of our action provides a tight constraint on the divergences in our calculations. The odd parity of the Chern-Simons field prevents any correction to the scalar field at the one-loop level: by rescaling $A_\mu \to g^{-1}A_\mu$, we see that under parity $g^2 \to -g^2$, and so correlations of ϕ which are parity invariant can only depend on g^4 and thus the first corrections are at two loops. Accordingly, we do not find any one-loop contributions to Z_ϕ or Z_μ. Further as mentioned previously, the Coleman-Hill Theorem shows that the β-function of the Chern-Simons gauge coupling receives no contribution beyond one-loop [43–46] in perturbation.

More simplification results from the vanishing of diagrams at any order with closed scalar loops: they necessarily introduce a combinatoric factor of N and do not provide corrections to the gauge field as $N \to 0$ and so a Maxwell term, $F_{\mu\nu}^2$, though allowed by symmetry, is not generated. The vanishing of diagrams with closed scalar loops was explicitly shown in the previous section [Eq. (8.39)]. The details of the combinatoric argument go as follows: consider a graph with only external gauge-field legs. Since an external gauge-field leg must necessarily connect to two internal ϕ legs (and possibly one internal gauge-field leg) and since the $U(N)$ index of ϕ is not carried by any of the other external legs, there must be a sum over that index. Since the sum is proportional to N, this graph vanishes as $N \to 0$. More complex internal topologies only *add* factors of N to the graph and do not change this result. Thus, $Z_A = 1$. For the same reason there exists only a handful of potential two-loops contributions to the renormalization functions Z_X.

All the non-vanishing two-loop Feynman diagrams are shown in Figs. 8.7 and 8.8. Figures 8.7(a) and 8.7(b) are contributions to Z_ϕ, while Figs. 8.7(c) through 8.7(f) are corrections to the cubic gauge vertex. Figure 8.7(g) is the same as Fig. 8.7(a) but evaluated at zero external momentum, contributing to Z_μ.

The evaluation of these graphs is straightforward, since their singularity structures involve only simple poles by power counting. To evaluate each two-loop diagram, we perform the first momentum integral in $d = 3$ and then use dimensional regularization on the remaining single integral. This

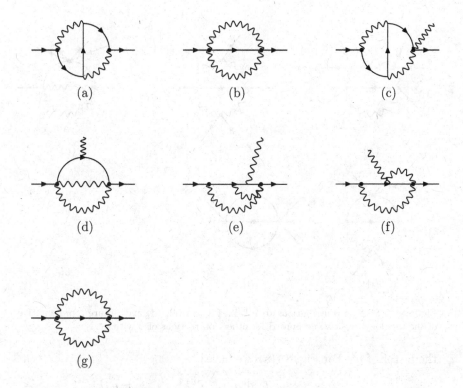

Fig. 8.7 Two-loop diagrams arising from the gauge field coupling to the scalar field: contributions to Z_ϕ (a), (b); corrections to the cubic gauge vertex (c)–(f); contributions to $Z_\mu(p = 0)$ (g).

scheme works because the first integral can only give power law divergences since any integrand with odd powers of momentum vanishes, leaving only even powers of the momenta compared to the three-dimensional measure.

And as previously mentioned, by power counting the remaining integral must diverge logarithmically and cannot cancel the power-law divergence from the first integration. Thus the logarithms only arise in the second of the two integrations. A more sophisticated treatment suggests that this scheme is consistent to all orders [42]. As is usual, we perform the necessary tensor algebra in physical dimensions before analytically continuing the dimensions of the resulting scalar integrand. Our results for each graph agree with those in [42].

For the scalar field wavefunction renormalization Z_ϕ, there are two divergent two-loop diagrams contributing. To extract Z_ϕ, we need to isolate

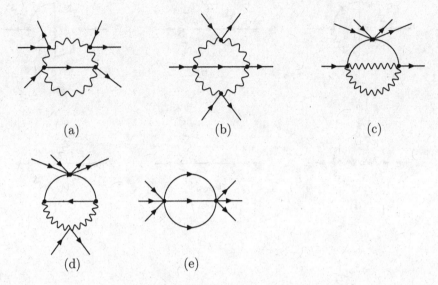

Fig. 8.8   Two-loop contributions to $\Gamma^{(6,0)}$. Figures (c), (d) and (e) are representative of the topology; we have accounted for other contractions of ϕ with ϕ^*.

the factor of p^2. For Fig. 8.7(a), we have

$$\text{Fig. 8.7(a)} = 16g^4 \int \frac{\mathrm{d}^d q}{(2\pi)^d} \int \frac{\mathrm{d}^d k}{(2\pi)^d} \frac{\epsilon_{\alpha A\beta}\, p_\alpha q_A k_\beta}{q^2 k^2 [(p-q)^2 + \mu]}$$
$$\times \frac{\epsilon_{\gamma B\delta}\, q_\delta k_B p_\delta}{[(p-q-k)^2 + \mu][(p-k)^2 + \mu]} \,. \tag{8.55}$$

The algebraic structure of the resulting expressions from the two-loop Feynman diagrams are considerably more complex than their one-loop counterparts. After performing the integration over k after using the mathematical identity for $d = 3$

$$k_\alpha k_\beta \rightarrow \frac{1}{d}\delta_{\alpha\beta} k^2 \,, \tag{8.56}$$

we are left with the integral

$$\text{Fig. 8.7(a)} = \frac{g^4}{2} p_\alpha p_\delta \delta_{\beta B} \int \frac{\mathrm{d}^d q}{(2\pi)^d} \frac{\epsilon_{\alpha A\beta}\, \epsilon_{\gamma B\delta}\, q_A q_\gamma}{q^4 |q|} \,. \tag{8.57}$$

By use of the following relation

$$\epsilon_{\alpha A\beta}\, \epsilon_{\gamma B\beta} = \delta_{\alpha\gamma}\delta_{AB} - \delta_{\alpha B}\delta_{A\gamma} \,, \tag{8.58}$$

and the method of Feynman parameters

$$\frac{1}{A_1^{m_1} A_2^{m_2} \cdots A_n^{m_n}} = \int_0^1 dx_1 \cdots dx_n \, \delta\left(\sum x_i - 1\right) \frac{\prod x_i^{m_i-1}}{[\sum x_i A_i]^{\sum m_i}}$$
$$\times \frac{\Gamma(m_1 + \cdots + m_n)}{\Gamma(m_1) \cdots \Gamma(m_n)}, \tag{8.59}$$

we arrive at the following result for the divergent part of Fig. 8.7(a):

$$\text{Fig. 8.7(a)} = -\frac{g^4}{6\pi^2 \epsilon} p^2. \tag{8.60}$$

The remaining diagrammatic contribution to Z_ϕ has the following expression:

$$\text{Fig. 8.7(b)} = -8g^4 \int \frac{d^d q}{(2\pi)^d} \int \frac{d^d k}{(2\pi)^d} \frac{q \cdot k}{q^2 \, k^2 \, [(p-k-q)^2 + \mu]^2}. \tag{8.61}$$

Introducing a Feynman parameter x and using Eq. (8.59), we are left with the following integral

$$-\frac{g^4}{\pi} \int_0^1 dx \frac{x}{\sqrt{x(x-1)}} \int \frac{d^d k}{(2\pi)^d} \frac{k \cdot (p-k)}{k^2 \sqrt{(p-k)^2 + \frac{\mu}{1-x}}}. \tag{8.62}$$

Introducing a second Feynman parameter a and going through the same procedure, we come to an integral that can be separated into two terms: a term proportional to p^2 and another term independent of the external momentum:

$$-\frac{g^4}{\pi} \int_1^0 da \frac{1}{2\sqrt{a}} \int \frac{d^d k}{(2\pi)^d} \left(\frac{a(a-1)p_A p_B}{(k^2 + \Delta)^{3/2}} - \frac{k_A k_B}{(k^2 + \Delta)^{3/2}} \right). \tag{8.63}$$

Only the first term contributes to Z_ϕ since it has the correct tensorial structure. Its evaluation leads to the following result for Fig. 8.7(b):

$$\text{Fig. 8.7(b)} = -\frac{g^4}{24\pi^2 \epsilon} p^2. \tag{8.64}$$

Incidentally, for Fig. 8.7(g) in which we are interested for its contribution to Z_μ, the external momentum $p \to 0$ limit is required. In this case, the term independent of p remains the only non-vanishing contribution, and it takes on the following value:

$$\text{Fig. 8.7(g)} = -\frac{\mu g^4}{8\pi^2 \epsilon}. \tag{8.65}$$

Table 8.1.   Divergent contribution from each graph and to the appropriate renormalization constant. Note that there are no divergences at one loop.

Graph	Divergence	Graph	Divergence	Graph	Divergence
1a	$-\dfrac{g^4 p^2}{6\pi^2 \epsilon}$	1e	$-\dfrac{g^5}{12\pi^2 \epsilon}$	2b	$\dfrac{3g^8}{4\pi^2 \epsilon}$
1b	$-\dfrac{g^4 p^2}{24\pi^2 \epsilon}$	1f	$\dfrac{g^5}{12\pi^2 \epsilon}$	2c	$\dfrac{15wg^4}{8\pi^2 \epsilon}$
1c	$-\dfrac{g^5}{6\pi^2 \epsilon}$	1g	$-\dfrac{\mu g^4}{8\pi^2 \epsilon}$	2d	$\dfrac{3wg^4}{2\pi^2 \epsilon}$
1d	$-\dfrac{g^5}{24\pi^2 \epsilon}$	2a	$\dfrac{g^8}{\pi^2 \epsilon}$	2e	$\dfrac{33w^2}{4\pi^2 \epsilon}$

Collecting all results together, we obtain the divergent parts of Z_ϕ, Z_μ, and Z_A:

$$Z_\phi = 1 - \frac{5g^4}{24\pi^2 \epsilon}, \tag{8.66}$$

$$Z_\mu = 1 - \frac{g^4}{8\pi^2 \epsilon}, \tag{8.67}$$

$$Z_A = 1. \tag{8.68}$$

Without presenting further cumbersome details, we list the divergent part of two-loop corrections to the cubic gauge vertex and the six-point w vertex in Table 8.1. They produce the divergent part of the relevant renormalization constants as follows:

$$Z'_g = 1 - \frac{5g^4}{24\pi^2 \epsilon}, \tag{8.69}$$

$$Z_w = 1 + \frac{7g^8}{4w\pi^2 \epsilon} + \frac{27g^4}{8\pi^2 \epsilon} + \frac{33w}{4\pi^2 \epsilon}, \tag{8.70}$$

from which we find that at two-loop order in $d = 3 - \epsilon$

$$\beta_g = -\frac{\epsilon}{2} g \tag{8.71}$$

$$\eta_\phi = \frac{5g^4}{12\pi^2} \tag{8.72}$$

$$\beta_\mu = \mu \left(-2 + \frac{g^4}{6\pi^2} \right) \tag{8.73}$$

$$\beta_w = -w \left(2\epsilon - \frac{7g^8}{2w\pi^2} - \frac{8g^4}{\pi^2} - \frac{33w}{2\pi^2} \right) = \frac{33}{2\pi^2}(w - w_+)(w - w_-) \tag{8.74}$$

where

$$w_\pm(g) = \frac{1}{33} \left[2\pi^2\epsilon - 8g^4 \pm \sqrt{4\pi^4\epsilon^2 - 32\pi^2 g^4\epsilon - 167g^8} \right] . \tag{8.75}$$

For nonvanishing g, as $\epsilon \to 0$ both roots are complex and there is no physical w fixed point. However, for $g^4 \leq 0.851\epsilon$, the radicand in (8.75) is real and w_+ is a stable fixed point for $\epsilon > 0$. Note, however, that for $\epsilon > 0$ the gauge coupling runs away to large values and thus the only stable fixed point in $d = 3 - \epsilon$ is at $(g, w) = (0, 4\pi^2\epsilon/33)$. Focusing on $d = 3$, we see that g is exactly marginal and $w_\pm = -g^4(0.242 \mp 0.392i)$. Writing $M = M_0 e^{-\ell}$, we have

$$w(\ell) = \Re w_+ + \Im w_+ \cot \left(\cot^{-1} \left[\frac{w_0 - \Re w_+}{\Im w_+} \right] + \kappa\ell \right) \tag{8.76}$$

where $\kappa = \frac{33}{2\pi^2 g^4} \Im w_+ = 0.655$. Note that this solution is unstable and as ℓ grows runs away to negative values of w, reminiscent of the behavior in a superconductor [47] and signaling a first-order transition. However, until any new fixed point controls the scaling, the critical exponents ν and η are determined entirely by the exactly marginal coupling g:

$$\eta = \frac{5g^4}{24\pi^2}, \tag{8.77}$$

$$\nu = \frac{1}{2 - g^4/6\pi^2} \approx \frac{1}{2} + \frac{g^4}{24\pi^2} . \tag{8.78}$$

We thus see that a chemical potential for writhe can alter the radius-of-gyration exponent ν and therefore writhe alters the universality class of a self-avoiding walk before driving it to collapse. The gauge field is not perturbatively renormalized and thereby preserves its topological character.

Though one might have expected that adding a bias for writhe would collapse the polymer so that $\nu < \frac{1}{2}$, we have found the contrary. However, imposing a bias on a self-avoiding walk constrains each step which in turn requires the polymer to swell. This is similar to the result that, on average, a long stretch of polymer is needed to even form a closed loop [48]. It is instructive to examine the behavior of the writhe in this system. As in [15] the scaling behavior of the average writhe $\langle Wr \rangle$ and the average squared writhe $\langle Wr^2 \rangle$ are given in terms of the specific heat exponent $\alpha = 2 - d\nu = \frac{1}{2} - \frac{g^4}{8\pi^2}$:

$$\langle Wr \rangle \sim -\frac{d}{d(g^2)} \ln \left(L^{\alpha-2} \right) = \frac{g^2}{4\pi^2} \ln L \tag{8.79}$$

$$\langle Wr^2 \rangle \sim \frac{d^2}{d(g^2)^2} \ln \left(L^{\alpha-2} \right) = -\frac{1}{4\pi^2} \ln L \tag{8.80}$$

we see that if the chemical potential vanishes then $\langle Wr \rangle = 0$ as expected. These logarithmic corrections are in addition to the writhe that is stored is polymer segments comparable to the persistence length which scales as L, a result which does not violate the rigorous bound $\ln\langle |Wr| \rangle \geq \frac{1}{2} \ln L$ [50, 51]. Since the mean-square writhe decreases, we see that the self-avoidance of the polymer, even at the Θ point, makes it more difficult to have contorted, writhing configurations. Our expressions of scaling exponents η and ν depend continuously on g^2 and are thus similar to those in the two-dimensional XY-model. There, topological defects are responsible for this uncommon behavior; here the topological link constraint is responsible.

8.6   Conclusion

In this chapter, we mapped the study of the statistics of self-avoiding random walks in a Θ-solvent to a Chern-Simons field theory and found a new scaling regime for topologically constrained polymers by calculating the scaling exponents η and ν to two-loop order at the Θ-point.

Since this global, topological constraint can alter the universality class of the walk we expect that other constraints such as knotting classes will also alter scaling behavior. We conjecture that even in a good solvent the scaling behavior will be altered, though as prior work indicates [15] it is difficult to establish results in a controlled approximation. Since taking $N \to 0$ amounts to canceling a functional determinant, progress in this problem might be made by introducing fermionic partners to the complex scalars to have the same effect. It is possible that supersymmetric formulation of this field theory would yield a more complete understanding of the effect of topological constraints. Note that above the Θ-point we may use the $U(N)$ ϕ^4-theory to describe the polymers. In this case we may introduce the auxiliary Hubbard-Stratonovich field χ [15] to find the resulting partition function as follows:

$$\mathcal{Z} = \int [d\chi][dA] \det \left[-|\partial - igA|^2 + \mu + i\sqrt{2u}\chi \right]^{-N}$$
$$\times \exp \left\{ -\frac{1}{2} \int d^3x \left[\chi^2 + \epsilon^{\mu\nu\rho} A_\mu \partial_\nu A_\rho \right] \right\}, \qquad (8.81)$$

an apparently, purely quadratic theory as $N \to 0$. The topological nature of the Chern-Simons field coupled to the supersymmetry breaking [49] $N \to 0$ limit deserves more attention in this context and in other related models.

References

[1] P. G. de Gennes, *Phys. Lett. A* **38**, 609 (1972).

[2] P. G. de Gennes, *J. Phys. (France) Lett.* **36**, 55 (1975).

[3] B. Duplantier, *J. Phys. (France)* **43**, 991 (1982).

[4] B. Duplantier, *J. Chem. Phys.* **86**, 4233 (1987).

[5] P. Grassberger, *Phys. Rev. E* **56**, 3682 (1997).

[6] C. W. Young, J. H. R. Clarke, J. J. Freire and M. Bishop, *J. Chem. Phys.* **105**, 9666 (1996).

[7] A. M. Rubio and J. J. Freire, *J. Chem. Phys.* **106**, 5638 (1997).

[8] J. H. White, *Amer. J. Math.* **91**, 693 (1969).

[9] F. B. Fuller, *Proc. Nat. Acad. Sci. USA*, **68**, 815 (1971).

[10] B. Fain and J Rudnick, *Phys. Rev. E* **60**, 7239 (1999).

[11] B. Fain, J. Rudnick and S. Ostlund, *Phys. Rev. E* **55**, 7364 (1997).

[12] F. Julicher, *Phys. Rev. E* **49**, 2429 (1994).

[13] J. F. Marko and E. D. Siggia, *Phys. Rev. E* **52**, 2912 (1995).

[14] T. Strick, J. Allemand, D Bensimon, A. Bensimon and V. Croquette, *Science* **271**, 1835 (1996).

[15] J. D. Moroz and R. D. Kamien, *Nucl. Phys. B* **506**, 695 (1997).

[16] P. M. Chaikin and T. C. Lubensky, *Principles of Condensed Matter Physics*, Cambridge University Press, New York, 1980.

[17] R. B. Griffths, *Phys. Rev. B* **7**, 545 (1973).

[18] I. D. Lawrye and S. Sarbach, *Phase Transitions and Critical Phenomena*, Vol. 9, Academic Press, London, 1983.

[19] M.G. de Amaral, *J. Phys. G* **24**, 1061 (1998).

[20] P. Arnold and D. Wright, *Phys. Rev. D* **55**, 6274 (1997).
[21] A. B. Zamolodchikov, *Sov. J. Nucl. Phys.* **44**, 529 (1986).

[22] J. Polchinsky, *String Theory*, Vol. 2, Cambridge University Press, Cambridge, 1999.

[23] D. Friedan, Z. Qiu and S. Shenkar, *Phys. Rev. Lett.* **52**, 1575 (1984).

[24] A. De Martino and M. Moricoru, *Nucl. Phys. B* **528**, 577 (1998).

[25] C. Itzykson and J. M. Drouffe, *Statistical Field Theory*, Vol. 1, Cambridge University Press, Cambridge, 1989.

[26] M. J. Stephen, *J. Phys. C* **13**, 283 (1980).

[27] F. Wegner and E. K. Riedel, *Phys. Rev. Lett.* **29**, 349 (1972).

[28] J. S. Hager and L. Schafer, *Phys. Rev. E* **60**, 2071 (1999).

[29] J. S. Hager, *J. Phys. A* **35**, 2703 (2002).

[30] S. C. Zhang, T. H. Hansson and S. Kivelson, *Phys. Rev. Lett.* **62**, 82 (1988).

[31] J. K. Jain, *Phys. Rev. Lett.* **63**, 199 (1989).

[32] S. C. Zhang, *Int. J. Mod. Phys. B* **6**, 25 (1992).

[33] G. Murthy and R. Shankar, preprint (1998), cond-mat/9802244.

[34] S. Deser, R. Jackiw and S. Templeton, *Phys. Rev. Lett.* **48**, 975 (1982).

[35] W. Siegel, *Nucl. Phys. B* **156**, 135 (1979).

[36] E. Witten, *Comm. Math. Phys.* **121**, 351 (1989).

[37] E. Efraty and V. P. Nair, *Phys. Rev. Lett.* **68**, 2891 (1992).

[38] A. D. Linde, *Phys. Lett. B* **96**, 289 (1980).

[39] K. Farakos, K. Kajantie, K. Rummukainen and M. Sharposhinikov, *Nucl. Phys. B* **425**, 67 (1994).

[40] C. Nash, *Differential Topology and Quantum Field Theory*, Academic Press, London, 1991.

[41] W. Chen, G. W. Semenoff and Y. S. Wu, *Phys. Rev. D* **46**, 5521 (1992).

[42] L. C. de Albuquerque, M. Gomes and A. J. da Silva, *Phys. Rev. D* **62**, 085005 (2000).

[43] G. W. Semenoff, P. Sodano and Y.S. Wu, *Phys. Rev. Lett.* **62**, 715 (1988).

[44] A.J. Niemi and G. W. Semenoff, *Phys. Rev. Lett.* **51**, 2077 (1983).

[45] A. N. Redlich, *Phys. Rev. D* **29**, 2366 (1984).

[46] S. Coleman and B. Hill, *Phys. Lett. B* **159**, 184 (1985).

[47] B. I. Halperin, T. C. Lubensky and S.-K. Ma, *Phys. Rev. Lett.* **32**, 292 (1974).

[48] P.-G. de Gennes, *Introduction to Polymer Dynamics*, Chap. 2, Cambridge University Press, Cambridge, 1990.

[49] S. Golowich and J. Z. Imbrie, *Commun. Math. Phys.* **168**, 265 (1995).

[50] E. Orlandini, M. C. Tesi, S. G. Whittington, D. W. Sumners and E. J. J. van Rensburg, *J. Phys. A: Math. Gen.* **27**, L333 (1994).

[51] E. J. J. van Rensburg, E. Orlandini, D. W. Sumners, M. C. Tesi and S. G. Whittington, *J. Phys. A: Math. Gen.* **26**, L981 (1993).

PART 5
Summary

Chapter 9

Final Thoughts

In this monograph, we have examined the thermodynamic and structural properties of charged and fuzzy colloids, as well as the critical properties of topologically constrained polymers. For the charged and fuzzy colloids, we use a foam model to map the interactions between colloidal particles to interactions between flat surfaces, and we account for the observation of non-close packed lattices in terms of the mathematical constraint of minimal surfaces. In addition to entropy, we have shown that the underlying repulsion of the charged and fuzzy colloidal crystals also contribute to the observed phase properties for these systems. Unlike the traditional methods such as density-functional theory and molecular dynamics (MD) simulations, our foam model is intrinsically many-body and thus represents a global approach to the modeling of these systems. Our foam model also has the advantage that we can compute directly the various elastic constants in the same formalism from first principles.

For the topologically constrained polymers, we have discovered that non-trivial topologies in the structure of long polymers indeed lead to a new universality class at critical points. To compute the fluctuation-induced correction to the radius-of-gyration exponent ν near the Θ-point, we consider the three-dimensional symmetric $U(N)$-Chern Simons theory in the $N \to 0$ limit. We have computed our results to two-loop order. We conjecture that even in a good solvent the scaling behavior will be altered by the introduction of topological constraints to the polymer configurations.

In the larger picture, the results in this monograph demonstrates that in many instances a heuristic analytical model that captures essential features of the system under study can offer tremendous value and utility in obtaining a qualitative big picture on the inner workings of a complex system. We see that phenomenology plays an essential role in the modeling of many complex systems for which first-principle computations often prove prohibitive. Of course, no computations and simulations can replace the importance of experimental verification in our continued search of further understanding of Nature.

While there has always been cross pollination of ideas between different branches of physics, it would be interesting to see if ideas developed in this monograph can find interesting applications in other fields

Index

β-function, 29, 34, 35, 160, 162, 164
β-function
  zeroes of, 35
β-tungsten, 89
ϵ-expansion, 148, 153, 159
$\epsilon_{\mu\nu\lambda}$, 162
$\mathcal{M}_m$ model, 149
ϕ^6 potential, 149
$\mathcal{SM}_n$, 149
$\mathbb{Z}_2$-symmetry, 22
$\mathbb{Z}_N$-symmetry, 22
Θ-point, 151, 154, 158, 159, 170, 177
Θ-solvent, 170
Θ-transition, 150

A15 lattice, 85
  $Pm\bar{3}n$, 85
  instability, 135, 136
abelian Chern-Simons theory, 13
abelian gauge group, 11
absorption of helium on
  krypton-plated graphite, 149
actin, 148
action, 164
algebraic structure, 166
algorithm, 5, 6
alkyl chain, 8, 49, 111
alkyl corona, 137
aluminum, 85
analytical close form, 5
analytical form, 5
analytical model, 5

anomalous dimension, 147, 162
antisymmetry, 161–163
anyons, 156
approximate analytical model, 4–6
arc length, 150, 158
Archimedean solid, 94
arithmetic, 84
aromatic ring, 8, 49, 137
associativity, 52
asymptotic flatness, 18
atom, 4
ATP, 158

backscattering, 117
Bain strain, 127
Banach space, 35
  nonempty convex subset of, 35
bare charge, 123
barium, 88
Barker-Henderson effective diameter, 131
Barker-Henderson perturbation, 117
basis, 30
  vectors, 53
BCC lattice instability, 127, 135, 136
bending mode, 43
bias, 169
big bang, 17, 18
bilayer, 91
biological macromolecule, 115
biology, 147
block spins, 27

body diagonal, 136
body-centered cubic lattice (BCC), 85
Bogoliubov's $\bar{R}$-operation, 153
Bohr, Neil, 4
Born, Max, 4
bosonic theory, 149
boundary group, 84
BPHZ renormalization scheme, 152
Bragg angle, 117
Bragg reflection, 117
brain, 17
Bravais lattice, 50, 85
Brouwer fixed-point theorem, 35
Brownian motion, 6
Brownian particle, 6
bulk density, 110
bulk free energy, 103, 104, 109, 111,
    112, 123, 127, 129, 130
bulk modulus, 39, 117, 123–125,
    129–131, 133, 134, 138, 139,
    141–143

calculus, 60, 77
Callan-Symanzik equation, 34, 153,
    160
cation, 110
cellular framework, 120
cellular free-volume theory, 104
centered cell, 55
centered rectangular lattice, 54
central charge, 149
ceramic precursor, 115
cesium, 88
chain connectivity, 43, 147, 148, 151
chained polymers, 23
charge density, 123
charge renormalization, 124
charged colloids, 8, 12, 13, 46, 49, 91,
    103–105, 112, 115–117, 120,
    122–124, 129–131, 137, 142, 177
charged plate, 110
chemical potential, 13, 39, 41, 42,
    158, 170
Chern-Simons coupling, 156
Chern-Simons gauge field, 13, 148,
    155, 158, 160, 163, 164

odd parity, 164
Chern-Simons theory, 148, 155–158,
    161, 170
    abelian, 156–159
    non-abelian, 156
chromium, 88
class, 77
classical analysis, 5
classical density-functional theory, 25
classical mechanics, 5
close packing, 129, 141
    limit, 133, 135, 143
closed analytical form, 5
closed loop, 152
closed scalar loops, 160
closest-packed volume fraction, 88
closure, 52
coarse grain, 33
coarse graining, 9, 10, 28, 29, 33, 34
coarse-grained models, 7, 9
coarsening, 7
coating materials, 115
coexistence curve, 119, 120
coexistence point, 119
coexistence region, 119
Coleman-Hill Theorem, 164
collective effect, 105
colloidal crystals, 11, 13
colloidal engineering, 143
colloids, 9, 38, 42, 49
Coloumb-energy functional, 40
columnar site, 90, 135, 136
compact topological space, 35
complex roots, 169
complexity, 4
computation tools, 6
configuration integral, 104
configuration space, 23, 46
conformal field theory, 149
continuity, 77, 80
continuous phase transitions, 21, 25
control parameter, 8
converging series, 5
coordinate system, 11
coordination number, 88
copper, 85

corona, 91, 111, 115, 118, 137
coronal brush, 112
correlation, 9, 39, 41, 104, 130, 147
  function, 10, 39, 152, 153, 155, 156,
    158–160
  n-th order, 44
  length, 26, 29–31, 33, 36, 148, 151
    exponent ν, 151
correspondence principle, 11
cosmic microwave background, 18
cosmology, 18
Coulomb interaction, 8, 38
counterion, 110, 123
coupling constant, 10
covariance coupling, 155
critical behavior, 159
critical dimension, 153
critical exponents, 10, 13, 26, 31, 32,
  34, 36, 169
critical opalescence, 21
critical phenomena, 10, 25, 149, 151,
  177
critical point, 10, 21, 25, 30, 32, 147,
  177
critical regime, 154
crossover, 133, 148
crystallographic notation, 78, 85, 98
  Hermann-Mauguin, 55
crystallographic restriction theorem,
  53
crystallography, 55
cube, 80
cubic elastic constants, 129
cubic gauge vertex, 164, 168
curvature, 78, 109, 128
cutoff, 33
cyclic group, 84
cylinder, 11, 50, 78

de Broglie, Louis, 4
de Broglie wavelength, 104
de Rham cohomology group, 85
Debye screening length, 109, 110, 133,
  137
Debye-Hückel approximation, 110,
  115

defect, 84
deformation, 80, 126, 131, 135
degrees of freedom, 39
dendrimer, 138
dendritic polymers, 90
density, 24
density-functional theory, 9, 12, 38,
  41, 43, 46, 177
  classical, 42, 45
  effective potential, 40
  electronic, 38, 42
  modified weighted, 44–46
  weighted, 43
Derjaguin approximation, 110, 117,
  133
dielectric constant, 109, 110, 122
diffeomorphism invariance, 156
differential form, 85
dihedral elements, 78
dimensional analysis, 152
dimensional regularization, 153, 159,
  164, 165
dimensionless electrostatic potential,
  109
dipole-dipole interaction, 20
Dirac, Paul, 4
direct correlation function, 45, 46
directed line segment, 84
discontinuity, 10
discreet symmetry, 22
  global, 22
discrete internal symmetry, 23
disorder, 27
disordered phase, 8, 116
dispersion forces, 130
divergence, 10, 148, 163, 168
DNA, 13, 148
  plasmid, 158
dodecahedron, 80
dodecyl chains, 138
duality, 85
dynamical system, 7
dynamics, 7

economics, 147
edge, 80, 82

Edwards model, 150
effective charge, 122, 124
effective description, 5
effective diameter, 131
effective models, 5, 7
effective partition function, 158
effective potential, 104
effective screening length, 131
effective surface charge, 122
Ehrenfest classification, 10, 23
    higher-order phase transitions, 24
eigen-expansion, 30
eigenvalue, 31, 32
Einstein crystal, 105
elastic constants, 124, 129–134,
    138–140, 142, 177
elastic energy, 129
elastic free energy, 13, 124–126
elastic properties, 13
elasticity, 11
    tensor, 124
electrolytes, 109
electromagnetism, 3, 4, 11, 18, 157
electron-electron Coulomb repulsion,
    41
electronic density, 39, 40
electrostatic potential, 109
electrostatics, 8, 110
electroweak, 18
    interaction, 156
    theory, 4
elongational shear, 125, 127
    mode, 125, 126, 135
emulsions, 38
energy scale, 10
engineering dimension, 153, 160
entropy, 43, 103, 120, 177
enzyme, 158
equilibrium, 7
    thermodynamics, 9, 42
equivalence relation, 77
equivalent classes, 77
Euclidean space, 57
Euler characteristic, 80, 84, 98
excess free energy, 45
excess volume, 91

exchange interaction, 39
exchange-correlation potential, 41
excluded volume, 147
    interaction, 38
    repulsion, 111
expectation, 151
    value, 156
exponential, 157
    interaction, 112
extermization, 9
external leg, 153, 161
external momentum, 164, 167

face, 80
face-centered cubic lattice (FCC), 85
    $Fm\bar{3}m$, 85
fermionic partners, 170
ferromagnetism, 23
feudalism, 3
Feynman amplitudes, 153
Feynman diagrams, 151–153, 163,
    164, 166
field correlation, 34
field lines, 157
field-theoretic language, 13
financial markets, 3
finite group, 57
first-order phase transition, 8, 10, 20,
    150, 169
first-order transitions, 36
    discontinuity in first derivatives, 24
fitting parameter, 122
fixed point, 30, 169
    infrared, 154
    theorems, 34
fixed-volume geometric constraint,
    111
Flory theory, 111
fluctuation, 13, 25, 154, 177
fluctuation-dissipation theorem, 7
foam analogy, 39, 113, 116, 124, 143
foam model, 11, 12, 49, 50, 91, 103,
    113, 116, 177
foams, 38, 49
Fokker-Planck equation, 6, 7
force constant, 123

force-extension curves, 148
four-point coupling, 152
four-point vertex function, 162
Fourier space, 44
fractional quantum Hall effect, 156
fractional statistics, 156
free energy, 9, 10, 12, 42
   discontinuity in first-derivative, 24
   Helmholz, 42
   Helmholz excess, 43
   non-analyticity, 10, 23
free volume, 104, 105, 125–127, 135
   theory, 38, 104, 105, 108, 130
friction, 6
frustration, 112
functional determinant, 170
fundamental models, 11
fundamental theories, 5
fuzzy colloids, 8, 12, 13, 46, 49, 91,
   103–105, 112, 115, 129, 137, 142,
   177

gallium, 3
gas, 19
gauge, 159
   field, 157, 169
   group, 11, 157
   invariance, 156, 159
   propagator, 161, 163
   symmetry, 11
Gauss linking number, 156, 157
Gauss' law, 110, 119
Gaussian distribution, 45
Gaussian measure, 157
Gaussian statistics, 147
general covariance, 11, 156, 157
general relativity, 11, 156
genus, 80
geoemetric group theory, 57
geometric independence, 82
geometry, 13, 46
germanium, 3
glide, 85
   plane (g), 55
   reflection, 53, 78
global property, 149

globule, 147
gluon, 4, 160
   self-energy, 160
gold, 85
Goldberg tetrakaidecahedron, 95
good solvent, 177
gravity, 5, 11, 18, 123
ground state, 39, 41
group, 50, 51
   definition, 50
   structure, 52
GUT epoch, 18

H^2-group, 78
hadrons, 4
Hamiltonian, 27, 151
hardcore, 8
   interaction, 130
   packing, 38
   repulsion, 149
Hartree approximation, 41
Hartree-energy functional, 41
Hartree-Fock theory, 39
Hartree term, 41
He^3–He^4 system of binary mixtures,
   149
Heisenberg model, 23
Heisenberg, Werner, 4
Hermann-Mauguin notation, 12, 55,
   56
Hermann-Mauguin symbol, 55
hexagonal lattice, 54
Higgs boson, 19
   vacuum expectation value, 19
Higgs model, 149
high-density regime, 138
high-temperature expansion, 151, 152
high-temperature Ising model, 151
Hilbert problem, 95
Hilbert space, 157
Hilbert, David, 95
Hohenberg-Kohn theorem, 39, 40
holonomy, 157
homeomorphism, 59, 80, 82
homotopy group, 84, 98
Hubbard-Stratonovich field, 170

hydrodynamics, 7
hydrogen bonding, 19, 20
hydrogen chloride (HCl), 119, 129
hyperplane, 82

ice, 19
    crystalline structures, 19
    cubic (I_c), 19
    hexagonal (I_h), 19
ideal free energy, 138
ideal gas, 42
identification map, 77
identity element, 34, 52
imperialism, 3
index of refraction, 21
industrial revolution, 3
infinite-curvature limit, 50, 118
infinite-order phase transitions, 24
inflation, 18
information technology, 3
infrared fixed points, 35
infrared-stable fixed point, 154
interaction energy, 40
interaction parameters, 10
interaction potential, 110
interfacial area, 103
interfacial free energy, 103, 109–112,
    127, 129
interfacial properties, 42
intermolecular potential, 8
internal energy, 7
interparticle potential, 8
interstitial site, 90, 105, 135
inverse element, 52
inversion, 52
ionic strength, 115, 117
ionization, 123
iridescence, 117
iron, 88
irrelevant coupling constants, 31, 32,
    36
Ising model, 17, 22, 152
    $\mathbb{Z}_2$-symmetry, 22
    antiferromagnetic, 22
    Ernst Ising, 22
    ferromagnetic, 22

Hamiltonian, 22
    Onsager solution, 27
    specific case of Potts model, 23
isometry, 52, 78
isoperimetric quotient, 94, 95, 97
isostructural transitions, 119
isotropic shear modulus, 122
isotropy, 125

Kelvin's problem, 12, 94, 95, 98, 112,
    118
Kelvin's tetrakadecahedron, 88
Kepler conjecture, 12, 95, 98, 112
kinetic energy, 20, 40
    functional, 40
kinetics, 7
knot theory, 156
knotting classes, 170
Kohn-Sham approach, 40, 41
Kohn-Sham equation, 41
Kossel lines, 117
Kosterlitz-Thouless transition, 24

Lagrange multiplier, 41
Lagrangian, 156, 157
    metric-independence, 156
lambda line, 150
Landau gauge, 159
Landau-Ginzburg description, 149,
    151
Landau-Ginzburg theory, 9
Langevin equation, 6
Laplace transform, 151
latent heat, 10, 36
    of fusion, 20
    of vaporization, 21
lattice
    A15, 8, 11–13, 85, 89, 95, 98, 105,
        111, 113, 125, 131–136,
        138–143
    BCC, 8, 11–13, 85, 88, 95, 98, 105,
        111, 113, 116, 117, 119, 120,
        125, 127–135, 137–143
    BCT, 127
    body-centered, 85

centered rectangular, 53
charged, 112
close-packed, 103
cubic, 124, 125, 136
face-centered, 85
FCC, 8, 11–13, 85, 95, 98, 105, 111,
    113, 116, 117, 119, 125, 127,
    131–135, 137–143
fuzzy, 112
hexagonal, 53, 85
monoclinic, 85
non-close packed, 103
oblique, 53
orthorhombic, 85
rectangular, 53
simple, 85
square, 53, 151, 152
tetragonal, 85
triclinic, 85, 124, 136
trigonal, 85
lattice model, 149
lattice phase
    inhomogeneity, 43
lattices, 21
Laurent series expansion, 153
lead, 85
length scale, 10
light propagation, 157
    in crystals, 93
light scattering, 116
Lindemann rule, 118
linear algebra, 77
linear operator, 30
linear superposition, 110, 117
linearization, 109, 120, 123
linking number, 13, 148, 154, 158
liquid, 19
    crystals, 11, 38, 42, 49, 85
local coordinate system, 77
local density approximation (LDA),
    41
local physics, 157
local property, 149
logarithmic correction, 170
logarithmic divergence, 163, 165
long-wavelength limit, 7

loop, 157
Lord Kelvin, 93
low-frequency limit, 7

Möbius strip, 78
macroscopic field variables, 7
magnetic spins, 21, 24
magnetism, 21
magnetization, 7
manifold, 58, 60, 77, 98, 155, 156
many-body problem, 12, 39, 41
many-body system, 103, 177
many-electron system, 12, 39–41
    Hamiltonian, 40
marginal coupling, 169
    constants, 31
materials, 4
mathematical map, 77
mathematical spaces, 58
matter, 4, 160
    self-energy, 160
Maxwell-Chern-Simons theory, 156
Maxwell kinetic term, 156
Maxwell's equations, 11
Maxwell term, 164
mean-field theories, 9, 10, 153
mean-square writhe, 170
measure, 58
melting curve, 118
melting density, 130
membranes, 38
meniscus, 21
mesogens, 11, 49
metallurgy, 3
metastable states, 17
methanol, 119, 122
method of Feynman parameters, 167
metric, 58, 98, 155, 157
    independence, 156
    space, 58, 60
micelle, 111, 138
microscopic interparticle interaction,
    109
minimal surfaces, 103, 177
minimal-area conjecture, 115, 118

minimal-area constraint, 111, 116, 118

mirror plane (m), 55

modern physics, 9

modified weighted density-functional theory (MWDA), 44–46

molecular biology, 8

molecular dynamics (MD) simulation, 8, 46, 116, 118, 129, 177

molecular isotropy, 49

molecular point group, 53

molecular simulation, 38

momentum scale, 10, 153, 159

momentum space, 33, 35

monoid, 34

monomer, 8, 149

monomer-monomer interaction, 150

Monte Carlo simulation, 7, 117

multi-electron wavefunction, 39

multicritical point, 148

mutually disjoint subsets, 77

natural elements, 3

Nature, 11

n-polytype
    regular, 95

n-th order direct correlation function, 44

Nb_3Al, 89

nearest-neighbor interaction, 22, 27, 118, 133

neighborhood, 59

neutron, 4

Newton law, 46

Newton's equation, 6

Newton's laws of motion, 5

nickel, 85

non-abelian gauge group, 11

non-analyticity, 10

non-close packed lattices, 177

non-close packed structures, 112, 118, 135

non-Riemanian geometry, 11

nucleus, 4

numerical methods, 4–7

numerical modeling, 5

numerical simulations, 5, 38

N-body problem, 5, 6

N-component Ising model, 151

N-state Potts model, 22
    Hamiltonian, 22

$N = 1$ superconformal unitary minimal series, 149

$N = 1$ supersymmetric theory, 149

n-vector model, 17, 23
    Hamiltonian, 23

oblique lattice, 54

observable, 156

octahedron, 96, 105

octehedral group, 94

one-body external potential, 40, 43

one-loop, 148
    correction, 161
    diagrams, 161
    gauge coupling, 160

$O(N)$ index, 153

$O(n)$-model, 23

$O(N)$-symmetric ϕ^6-model, 148, 153

$O(N)$-symmetric ϕ^6-theory, 154

Onsager, Lars, 22

open disc, 60

open set, 60

open system, 42

optical tweezer, 124

orbifold, 55, 57, 78, 84, 98
    notation, 12, 55, 98

order parameter, 9, 22, 25, 26
    dimensionality of, 22

order-disorder transition, 117

orthic tetrakaidecahedron, 94, 95, 105

overcharging, 130

packing efficiency, 88

packing fraction, 120, 136

packing geometry, 85

pairwise interaction, 104

parameter space, 29, 34

parity invariance, 164

particle physics, 11

partition function, 28, 151, 158, 171

Pauli, Wolfgang, 4

pentagonal dodecahedron, 95
percolation, 23
persistence length, 147, 170
perturbation, 7, 34, 35, 161
   theory, 118
phase, 157
   diagram, 13, 119, 120
   property, 13, 177
   stability, 7, 9, 12, 91
   transition, 12, 13, 17, 30
      Θ, 150
      big bang, 17
      brain, 17
      critical point, 21
      electroweak, 18, 149
      ferromagnetic-paramagnetic,
         10
      first-order, 10–13, 20, 21, 169
      in theta solvents, 148
      magnetism, 17
      second-order, 10–12
         continuous, 21
   spins, 17
   water, 17
Phelan, Robert, 95
phenomenological models, 5, 10, 21
phenomenology, 5, 9, 177
phonon, 105
photon, 4
Planck epoch, 18
Planck, Max, 4
plane group, 53
Plateau border, 94
Plateau rules, 94
Plateau, Joseph, 94
point group, 52, 85
point particles, 50
Poisson-Boltzmann equation, 109, 120, 123
polaron, 149
polygon, 82, 95
   regular, 95
polyhedra, 80, 82, 94, 95
   regular, 95, 96
polymer, 9, 13, 38, 42

$O(N)$-symmetric ϕ^4-theory,
   $N \to 0$, 147
bending stiffness, 158
brushes, 111
chain, 150
   backbone, 158
closed, 158
   loop, 11–13, 148, 156
collapse, 169
configurations of, 177
conformation, 148, 158
critical behavior, 23, 150
critical exponents, 23
de Gennes' mapping, 147, 160
dentritic, 105
Edwards model, 150
free, 148, 154
   energy, 158
in solution, 43
linear, 8
linking number, 155
long, 12, 177
long-chained, 8
molecular structure, 149
monomer, 149
monomer-monomer interaction, 147
physics, 147
self-avoidance, 170
statistics, 158
topological constraints, 148
topologically constrained, 8, 12, 13, 23, 34–36, 170, 177
twist rigidity, 158
polymer chain interaction energy, 151
polystyrene sphere, 8, 116, 119
polystyrene suspensions, 117
polytope, 95
position, 6
potassium, 88
potential energy, 20
power counting, 163, 164
power-law divergence, 165
primitive cell, 55, 88
primitive translation vectors, 50, 52
primitive unit cell, 50

primitive vector, 90
probability function, 6
proton, 4

quantization, 156
quantum chromodynamics (QCD), 4, 34, 156
quantum electrodynamics (QED), 4, 33
quantum exchange correlation potential, 41
quantum field theory, 156
quantum gravity, 156
quantum mechanics, 3, 4, 39
quantum phase transitions, 24
quarks, 4
quotient space, 57, 77, 84

radicand, 169
radioactivity, 4
radius of gyration, 13, 147, 154
    exponent ν, 147, 154, 164, 169, 177
random walk, 147, 150, 151, 160
r-complex, 82
real space, 33
reciprocal lattice, 45, 88
    vectors, 24, 25, 45
rectangular lattice, 54
reduced density, 105, 108
reduced temperature, 31
reductionist approach, 6
reentrant transition, 118
reflection, 52, 53, 78, 125
    axis, 55
reflexivity, 77
regularization, 153
relevant coupling constants, 31, 32
renormalizable models, 28
renormalization, 153, 161
    constants, 168
    function, 153
    group, 9, 10, 12, 13, 23, 26, 30, 32, 34–36, 147, 158
        flow, 10, 29–31
            irrelevant coupling constants, 31

marginal coupling constants, 31
relevant coupling constants, 31
integration of momentum shell, 33
logarithmic correction, 148, 152
scale, 153, 160
scheme, 160
renormalized charge, 123
renormalized effective charge, 123
renormalized free energy, 152, 159
resonance, 123
response function, 39
rhombic dodecahedron, 87, 95, 105
rods, 50
rotation, 52, 53, 78
rotational symmetry, 125
running coupling constants, 29, 35, 169

S^2 group, 78
salt concentration, 8, 123
scalar ϕ^6-theory, 148, 149, 152, 158, 159
scalar field, 165
scale invariance, 25, 36
scaling, 10, 25, 26, 31, 169, 170
    argument, 122
    behavior, 13, 170, 177
    exponent, 147, 154, 170
    regime, 8, 147, 170
Schauder fixed-point theorem, 35
Schläfli symbols, 94, 96
Schrödinger, Erwin, 4
screened Coulomb interaction, 8, 110, 115–117, 130, 131
screened Coulomb potential, 109, 115
screening, 8
    length, 123
screw axis, 85
second moment, 151
second-order phase transitions, 8, 10, 21, 25, 33, 148, 150, 152
    loss of details, 33

second-order transitions, 10, 36
  discontinuity in second derivatives, 24
self-assembled micelles, 90, 105
self-avoiding random walk, 23, 151, 152, 169, 170
self-energy, 160
semi-group, 33, 34
semiconductors, 3
set theory, 59
sets, 59
shape anisotropy, 11
shear instability, 13, 135, 136
shear mode, 135
shear modulus, 105, 115–117, 120, 122–125, 127, 129–131, 135, 141–143
short-range interaction, 8
short-range repulsive potential, 49
silicon, 3
silver, 85
simple pole, 164
simple shear, 125
simplex, 80
  oriented, 84
  unoriented, 84
simplicial complex, 82
singularity, 164
six-point vertex function, 162, 163, 168
Slavnov-Taylor identities, 159
Smoluchowski equation, 7, 124
smoothness, 60, 77, 98
soap films, 94
soap froth, 118
sodium, 88
soft materials, 38
soft modes, 105, 131
soft repulsion, 103
solid, 19
  geometry, 12
  phase, 39
soliton, 149
solution conductivity, 123
solvation, 42
solvent, 149

good, 147, 150
poor, 147, 150
theta, 147, 150
space dimensionality, 22
space group, 53, 57, 85, 97
space-filling structure, 93
special relativity, 4
spectroscopy, 4
sphere, 11, 49, 80
spin, 151
  alignment, 27
  blocks, 27, 29
    effective interaction J', 28
    effective spin, 28
spin lattice, 9, 151
spin statistics, 156
spins, 17, 21, 152
spontaneous symmetry breaking, 19
square disc, 77, 78
square lattice, 54
standard model, 4, 11
static structure factor, 46
statics, 7
statistical mechanics, 9, 10, 42
statistics, 170
  transmutation, 156
stochastic differential equation, 6
strain, 125–127, 129
  tensor, 127
stretching mode, 43
strong interaction, 4
strong nuclear force, 11, 18
strontium, 85
structure, 9
  factor, 39
  function, 24, 25
Sundman theorem, 5
superconducting vortex, 85
superconductor, 9
superconformal minimal model, 149
superfields, 149
supersymmetry, 170
  breaking, 171
support
  convex, 119
surface acidic group, 123

surface charge density, 124
surface free energy, 129, 138
surface polymer coating, 38
surface potential, 110, 116, 119, 120,
  122, 129, 131
$SU(2)$, 11
$SU(2)$-gauge, 149
$SU(3)$, 11
symmetry, 9, 21, 77
  breaking, 20
  elements, 55
  factor, 153
  group, 97
  principle, 11

Taylor expansion, 25
temperature, 7
tensor algebra, 159, 165
tensorial structure, 167
tessellation, 57, 78, 82, 92, 98
tetrahedral angle, 94
tetrahedron, 80
theoretical model, 4
thermal fluctuation, 13
thermodynamic limit, 9
thermodynamic properties, 38
thermodynamics, 3, 7, 9–11, 152
theta point, 148, 150
theta solution, 34–36, 150
theta solvent, 147
theta transition, 147
three-body interaction, 151
three-body system, 5
Ti_3Sb, 89
time dependence, 7
time evolution, 6
titratable charge, 122
titration, 123
topoisomerase, 158
topological constraint, 148, 157, 158,
  170, 177
topological defect, 170
topological equivalence, 80
topological invariant, 156
topological quantum field theory, 156,
  158

topological space, 59, 77, 82, 84, 98
  definition of, 60
topology, 12, 57, 59, 60, 77
torus, 78, 80
trace, 157
trajectory, 6
  of motion, 5
transition probability matrix, 124
transitivity, 77
translation, 51, 53, 78, 85
  axis, 55
  vectors, 52
trial orbitals, 41
triangulation, 80, 82
triclinic derivative lattice, 136
tricritical Ising model, 149
tricritical phenomenon, 148, 149
tricritical point, 149–152, 159
triple point, 150
truncated octahedron, 94
truncation, 43
twist, 148, 156, 158
  rigidity, 148
two-body interaction, 151
two-dimensional lattice, 51
two-loop corrections, 164

ultraviolet fixed points, 35
unified force, 18
uniform electron gas, 41
unit ball, 35
unit cell, 50, 55
  centered, 55
  primitive, 55
unitary minimal model, 149
universal class, 12
universal functional, 40
universal solvent, 20
universality, 10, 24–26, 32, 36
  class, 8, 10, 13, 26, 169, 170, 177
universe, 9
$U(1)$, 148, 157
$U(1)$ gauge field, 11
$U(N)$ ϕ^4 theory, 170
$U(N)$-Chern Simons theory, 148, 164,
  177

$N \to 0$, 148
$U(N)$-index, 161, 164

vacuum, 157
    energy, 19
valence, 109
van der Waals, 43
    attraction, 130
    interaction, 20, 117, 149
vapor, 19
variational principle, 39, 118
velocity, 6
vertex, 80, 82
    function, 153, 159, 161
vibrational mode, 43
volume fraction, 115, 119, 120, 122,
    129
Voronoi cell, 49, 91, 103–105, 111

wallpaper group, 53, 57, 78, 82, 84, 97
    $c1m1$, 67
    $c2mm$, 55, 56, 68, 78
    $p1$, 62
    $p1g1$, 58, 78
    $p1m1$, 63
    $p2$, 61
    $p2gg$, 66
    $p2mg$, 65
    $p2mm$, 64
    $p3$, 72
    $p31m$, 74
    $p3m1$, 73
    $p4$, 69
    $p4gm$, 71
    $p4mm$, 70
    $p6$, 75
    $p6mm$, 76
    $p1g1$, 56

Ward identities, 159
water, 17, 119, 122
wavefunction renormalization, 165
Weaire, Denis, 95
Weaire-Phelan minimal surface, 90,
    105
weak nuclear force, 4, 11, 18
weighted density, 44
weighted density-functional theory, 43
weighting function, 44
Weinberg-Salem model, 33
Wigner's puzzle, 9
Wigner-Seitz approximation, 123
Wigner-Seitz cell, 49, 50, 88, 90, 104,
    123
    construction of, 49
Wilson lines, 156
Wilson loops, 155–158
Wilson, Kenneth, 26, 33
Wilson-Fisher fixed point, 147, 148,
    154
World Wide Web, 3
writhe, 13, 148, 157, 158, 169, 170

X-ray scattering, 119
XY-model, 23, 24, 170

Yang-Mills gauge theory, 157
Yukawa interaction, 129
Yukawa potential, 8, 116, 123

Z_A, 168
Z_μ, 164, 167, 168
Z_ϕ, 164, 165, 167, 168